MAPPED VECTOR BASIS FUNCTIONS FOR ELECTROMAGNETIC INTEGRAL EQUATIONS

© Springer Nature Switzerland AG 2022
Reprint of original edition © Morgan & Claypool 2006

Mapped Vector Basis Functions for Electromagnetic Integral Equations
Andrew F. Peterson

ISBN: 978-3-031-00558-9 Peterson, Mapped Vector Basis Functions for Electromagnetic Integral
    Equations (paperback)

ISBN: 978-3-031-01686-8 Peterson, Mapped Vector Basis Functions for Electromagnetic Integral
    Equations (e-book)

Library of Congress Cataloging-in-Publication Data

First Edition
10 9 8 7 6 5 4 3 2 1

MAPPED VECTOR BASIS FUNCTIONS FOR ELECTROMAGNETIC INTEGRAL EQUATIONS

Andrew F. Peterson
School of Electrical and Computer Engineering
Georgia Institute of Technology

ABSTRACT

The method-of-moments solution of the electric field and magnetic field integral equations (EFIE and MFIE) is extended to conducting objects modeled with curved cells. These techniques are important for electromagnetic scattering, antenna, radar signature, and wireless communication applications. Vector basis functions of the divergence-conforming and curl-conforming types are explained, and specific interpolatory and hierarchical basis functions are reviewed. Procedures for mapping these basis functions from a reference domain to a curved cell, while preserving the desired continuity properties on curved cells, are discussed in detail. For illustration, results are presented for examples that employ divergence-conforming basis functions with the EFIE and curl-conforming basis functions with the MFIE. The intended audience includes electromagnetic engineers with some previous familiarity with numerical techniques.

KEYWORDS

boundary element method, computational electromagnetics, integral equations, method of moments, parametric mapping

Contents

Preface

The discipline of computational electromagnetics has matured to the extent that there are now a number of commercial products in widespread use, and these tools are perceived to be useful to the community. In recent years, research has shifted from the basic foundations to specialized techniques, such as the various "fast methods" that accelerate the matrix solution process. Another research topic, which forms the focus of this monograph, is the use of higher order basis functions and curved cells to enhance the accuracy of the results. These techniques have not significantly penetrated the commercial marketplace.

My understanding of this subject has been largely influenced by two publications. The 1980 paper *Mixed finite elements in R3* by Jean-Claude Nedelec provided a comprehensive introduction to the two classes of vector basis functions considered in Chapters 3 and 4. The 1988 Ph.D. dissertation *Mixed-order Covariant Projection Finite Elements for Vector Fields* of Christopher W. Crowley contained a fairly complete discussion of the mapping procedures (for both divergence-conforming and curl-conforming functions) that I have expanded into Chapter 5. I am grateful for the efforts of these authors. Their papers arose from work directed toward the numerical solution of differential equations; it is hoped that the present text helps disseminate these ideas into the integral equation community. I am also grateful for the suggestions of John Shaeffer and Malcolm Bibby, who read an early draft of this manuscript.

A question remains to be answered: Are higher order techniques "better" than the status quo? At present, it is clear that high-accuracy results cannot be realized with low-order methods. It is also apparent that enhanced accuracy permits error estimation, improved confidence in numerical solutions, and the potential for "dialable" accuracy in the future. On the other hand, high accuracy will also require better representations for currents at all edges, and it will not prove meaningful without a commensurate improvement in the resolution of surface geometries and material parameters. As to which approach will prove best in the long run, the answer will be left to the user community.

CHAPTER 1

Introduction

Electromagnetics problems have been formulated in terms of integral equations since Pocklington analyzed wave propagation along wires in 1897, although the numerical solution of these equations (at least in the modern sense) is a relatively recent development. Rather than review the formulation and solution of these equations in detail, we refer the reader to a number of texts encompassing the scope of this discipline [1–6]. It is interesting that relatively sophisticated numerical techniques employing curved subsectional cells appeared more than three decades ago [7], but are not widely used even today. (Nor are they discussed in the standard texts [1–5] on the subject, except for a brief summary in [6].) One reason for this is that the majority of approaches in use rely on low-order polynomial basis functions, negating some of the accuracy advantages of curved cells. A second reason is that the implementation of curved cells in the full three-dimensional vector case involves a number of subtleties.

The present monograph attempts to walk the reader through the details of a curved-cell based discretization process. Our intended audience includes researchers and practitioners having some familiarity with flat-cell techniques. The scope of the text is restricted to perfectly conducting objects in three-dimensional space, with the surface of the object represented by a collection of contiguous patches. Wire structures are not considered, although the readers should be able to specialize these ideas to wires (or extend them to penetrable objects) without difficulty.

The remainder of Chapter 1 briefly reviews the formulation of two types of integral equation in widespread use for conducting bodies: the electric-field integral equation (EFIE) and the magnetic field integral equation (MFIE). The numerical discretization of these equations by the method of moments is also reviewed. Chapter 2 discusses several ways of representing curved-cell surface models. Chapters 3 and 4 provide motivation for the two types of vector basis functions in use for representing the surface current density within the integral equations, and reviews several families of these functions. Chapter 5 describes the manner in which those basis functions may be mapped to curved cells, while maintaining the critical continuity properties of the basis functions. Chapters 6 and 7 illustrate the complete numerical solution process, by applying the techniques to the EFIE and MFIE, respectively. A few numerical results are incorporated. A sample computer program that implements the EFIE approach is available as a companion to this text.

1.1   INTEGRAL EQUATIONS

Consider a perfect electric conducting body in an infinite homogeneous environment of permittivity ε and permeability μ. The body is illuminated by a sinusoidal steady-state source of electromagnetic radiation, having radian frequency ω, and all the time-varying quantities are represented by phasors with suppressed time dependence $e^{j\omega t}$. In the absence of the body, the source would produce an electric field $\bar{E}^{\mathrm{inc}}(x, y, z)$ and a magnetic field $\bar{H}^{\mathrm{inc}}(x, y, z)$ throughout the surrounding space. (These are denoted the *incident* fields.) In the presence of the conducting body, the fields are perturbed from these to the *total* fields $\bar{E}^{\mathrm{tot}}(x, y, z)$ and $\bar{H}^{\mathrm{tot}}(x, y, z)$. The perturbation can be accounted for by the presence of equivalent induced currents on the body.

The induced surface current density can be expressed as $\bar{J}(s, t)$, where s and t are parametric variables on the surface. [Henceforth, all references to the variable t are intended to denote position, not time.] The current density, if treated as a source function that exists in the absence of the conducting body, produces the *scattered* fields $\bar{E}^s(x, y, z)$ and $\bar{H}^s(x, y, z)$. The various fields are related by

$$\bar{E}^{\text{inc}} + \bar{E}^s = \bar{E}^{\text{tot}} \tag{1.1}$$

$$\bar{H}^{\text{inc}} + \bar{H}^s = \bar{H}^{\text{tot}} \tag{1.2}$$

Since the scattered fields are produced in infinite homogeneous space by $\bar{J}$, they are readily determined using any of the standard source-field relations,[1] such as

$$\bar{E}^s = \frac{\nabla(\nabla \bullet \bar{A}) + k^2 \bar{A}}{j\omega\varepsilon} \tag{1.3}$$

$$\bar{H}^s = \nabla \times \bar{A} \tag{1.4}$$

where the wavenumber of the medium is given by $k = \omega\sqrt{\mu\varepsilon}$, and the magnetic vector potential function is

$$\bar{A}(x, y, z) = \iint\limits_{\text{surface}} \bar{J}(s', t') \frac{e^{-jkR}}{4\pi R} ds' dt' \tag{1.5}$$

In (1.5), R is the distance from a point (s', t') on the surface to the point (x, y, z) where the field is evaluated. Throughout the text, primed coordinates will be used to describe the "source" of the electromagnetic field (the current density), while unprimed coordinates denote the "observer" location where that field is evaluated.

The total fields in (1.1) and (1.2) must satisfy the electromagnetic boundary conditions on the surface of the perfect electric body:

$$\hat{n} \times \bar{E}^{\text{tot}} \big|_{\text{surface}} = 0 \tag{1.6}$$

$$\hat{n} \times \bar{H}^{\text{tot}} \big|_{\text{surface}} = \bar{J} \tag{1.7}$$

where $\hat{n}$ is the outward normal unit vector, and in (1.7) the surface is approached from the exterior. By combining (1.1), (1.3), and (1.6), we obtain

$$\hat{n} \times \bar{E}^{\text{inc}} \big|_{\text{surface}} = -\hat{n} \times \frac{\nabla(\nabla \bullet \bar{A}) + k^2 \bar{A}}{j\omega\varepsilon}\bigg|_{\text{surface}} \tag{1.8}$$

---

[1] As in [6], (1.4) is employed as an alternative to the more traditional $\bar{B} = \nabla \times \bar{A}$.

This equation is one form of the *electric field integral equation*. (While it is in fact an integro-differential equation, the term "integral equation" is commonly used for simplicity.) The equation can be solved in principle for the surface current density $\bar{J}$ appearing within the magnetic vector potential $\bar{A}$.

By combining (1.2), (1.4), and (1.7), we obtain the *magnetic field integral equation*

$$\hat{n} \times \bar{H}^{\text{inc}}\,|_{\text{surface}} = \bar{J} - \hat{n} \times (\nabla \times \bar{A})|_{\text{surface}} \qquad (1.9)$$

Equation (1.9) is enforced in the limiting case from the exterior of the surface of the body. This equation may also be solved for $\bar{J}$. Once $\bar{J}$ is determined by a solution of (1.8) or (1.9), the fields and other observable quantities associated with the electromagnetic scattering problem can be determined by direct calculation.

The MFIE in (1.9) is only applicable to closed conducting bodies where (1.7) is a valid boundary condition, but the EFIE can be used for thin shells as well as solid structures where (1.6) holds. This restriction on the MFIE is due to the fact that (1.7) has been specialized to the situation where the magnetic field is zero on the inner side of the surface. The form of the MFIE in (1.9) is not valid if the surface is open and the field is nonzero on both sides of the surface.

Both equations can fail for closed bodies whose surfaces coincide with resonant cavities. In that situation, neither (1.6) nor (1.7) alone is sufficient to guarantee that the surface current density is the proper exterior solution. However, the two equations can be combined together to form the *combined field integral equation*. In this manner, the two boundary conditions are both invoked, and are sufficient to produce the proper solution. Reference [6] contains a discussion of this issue, known as the *interior resonance* problem, as well as the combined field equation and alternative remedies.

There are other integral equations that describe electromagnetic fields [8]; the scope here is limited to the EFIE and MFIE. Surface integral equations for application to dielectric bodies involve integral operators similar in form to (1.8) and (1.9). The mapping procedures used with Eqs. (1.8) and (1.9) can easily be extended to other equations.

1.2 THE METHOD OF MOMENTS

The vector integral equations in (1.8) and (1.9) have the form

$$L\{\bar{J}\} = \bar{V} \tag{1.10}$$

where L denotes a linear vector operator, $\bar{V}$ the given excitation function, and the equality holds for tangential vector components over the scatterer surface. To solve (1.10) numerically, it must be projected from the continuous infinite-dimensional space to a finite dimensional subspace (or in less mathematical terms, converted into a matrix equation). This process is known as *discretization*. One procedure for accomplishing this discretization is known as the *method of moments* (MoM) [1].

Our implementation of the MoM procedure involves approximating the quantity to be determined, $\bar{J}$, by an expansion in linearly independent vector basis functions

$$\bar{J}(s, t) \cong \sum_{n=1}^{N} I_n \bar{B}_n(s, t) \tag{1.11}$$

The N coefficients $\{I_n\}$ in (1.11) become the unknowns to be determined. To obtain N linearly independent equations, both sides of (1.10) are multiplied (using a scalar or dot product) with suitable vector testing functions $\{\bar{T}_m(s, t)\}$, and integrated over the surface of the body, to obtain

$$\iint_{\text{surface}} \bar{T}_m \bullet L\{\bar{J}\} ds\, dt = \iint_{\text{surface}} \bar{T}_m \bullet \bar{V}\, ds\, dt, \quad m = 1, 2, \ldots, N \tag{1.12}$$

By enforcing the equation in this manner, the boundary condition imbedded in the integral equation (1.10) is imposed throughout the domain of the testing function.

The equations in (1.12) can be organized into a matrix of the form $\mathbf{AI} = \mathbf{V}$, where the $N \times N$ matrix $\mathbf{A}$ has entries

$$A_{mn} = \iint\limits_{\text{surface}} \bar{T}_m \bullet L\{\bar{B}_n\} ds \, dt \tag{1.13}$$

and the $N \times 1$ column vector $\mathbf{V}$ has entries

$$V_m = \iint\limits_{\text{surface}} \bar{T}_m \bullet \bar{V} \, ds \, dt \tag{1.14}$$

The numerical solution of the matrix equation yields the coefficients $\{I_n\}$.

The method of moments is one of several general-purpose discretization procedures; it is closely related to the *weighted residual* method, the *Rayleigh Ritz* method, and the *finite element* or *boundary element* methods. Readers may consult [1–6] for specific examples of its use with a variety of scalar and vector electromagnetics equations, and discussions of the accuracy of the approximate results. The procedure can obviously be generalized in many ways from the specific implementation outlined above—for instance, more testing functions than basis functions can be employed to yield an overdetermined system [9]. Since such aspects are beyond the scope of the present text, we limit our consideration to the above.

As expressed in (1.11)–(1.14), the basis and testing functions may be spread over the entire surface, or defined with a more limited domain of support. The use of entire-domain basis functions, which are spread over the entire surface, is usually limited to codes developed for specific geometries. Because they allow flexibility in the geometry, codes based on subsectional functions are applicable to a broad class of problems. Henceforth, we restrict our attention to surfaces that are divided into a relatively large number of cells, and basis and testing functions with their domain limited to one or two adjacent cells each. Chapter 2 describes subsectional surface representations.

The choice of suitable basis and testing functions is the primary task before the user of the MoM procedure. These functions must be selected with a specific operator $L\{\bar{J}\}$ in mind. For instance, constraints on the continuity of the representation in (1.11) are dictated by the operator and usually play a large role in the choice of $\{\bar{B}_n\}$. Chapters 3 and 4 summarize two classes of vector basis functions known respectively as *divergence-conforming* functions and *curl-conforming* functions. Divergence-conforming functions maintain normal-vector continuity from cell to cell on the surface, while curl-conforming functions maintain tangential-vector continuity between adjacent cells. These basis functions are easily defined in square or triangular cells. To implement more general cell shapes, the basis function definition must be translated to curved cells, while maintaining the appropriate type of continuity between adjacent cells. This procedure is what we call *mapping*. Because the type of cell-to-cell continuity is different for divergence-conforming and curl-conforming functions, the type of mapping (Chapter 5) is also different. Since the EFIE involves a divergence operator, it is natural to employ divergence-conforming basis functions for its discretization. In contrast, the MFIE incorporates a curl operation, which suggests the use of curl-conforming basis functions. Chapters 6 and 7 provide specific details for curved-cell discretizations of the EFIE and MFIE.

REFERENCES

[1]   R. F. Harrington, *Field Computation by Moment Methods*. New York: IEEE Press, 1993. (Reprint of the original 1968 edition.)

[2]   R. Mittra, Ed., *Computer Techniques for Electromagnetics*. New York: Hemisphere, 1987. (Reprint of the original 1973 edition.)

[3]   N. Morita, N. Kumagai, and J. R. Mautz, *Integral Equation Methods for Electromagnetics*. Boston: Artech House, 1990. (Translation of the original 1987 edition.)

[4]   J. J. H. Wang, *Generalized Moment Methods in Electromagnetics—Formulation and Solution of Integral Equations*. New York: Wiley, 1991.

[5]   K. Umashankar and A. Taflove, *Computational Electromagnetics*. Boston: Artech House, 1993.

[6] A. F. Peterson, S. L. Ray, and R. Mittra, *Computational Methods for Electromagnetics*. New York: IEEE Press, 1998.

[7] D. L. Knepp and J. Goldhirsh, "Numerical analysis of electromagnetic radiation properties of smooth conducting bodies of arbitrary shape," *IEEE Trans. Antennas Propagat.*, vol. AP-20, pp. 383–388, 1972. doi:10.1109/TAP.1972.1140210

[8] H. Contopanagos, B. Dembart, M. Epton, J. J. Ottusch, V. Rokhlin, J. L. Visher, and S. M. Wandzura, "Well-conditioned boundary integral equations for three-dimensional electromagnetic scattering," *IEEE Trans. Antennas Propagat.*, vol. 50, pp. 1824–1830, 2002. doi:10.1109/TAP.2002.803956

[9] M. M. Bibby and A. F. Peterson, "On the use of over-determined systems in the adaptive numerical solution of integral equations," *IEEE Trans. Antennas Propagat.*, vol. 53, pp. 2267–2273, July 2005. doi:10.1109/TAP.2005.850729

C H A P T E R 2

The Surface Model

The numerical representation of an object's surface by an idealized subsectional model determines how well that surface can be approximated—how smooth it is, its precise curvature, its actual location, etc. Any subsequent analysis is limited by the accuracy of the surface representation. Furthermore, a computer code used for electromagnetic analysis is usually tightly coupled to the specific type of representation: the number of nodes and degree of polynomial, for instance, used to describe each "patch" or subsection of the surface. In this chapter, the basic framework for modeling surfaces on a cell-by-cell basis is reviewed.

Scalar mappings to define curved cell shapes are widely used in connection with the finite element solution of differential equations [1–4]. A cell-by-cell parametric mapping was used in connection with the numerical solution of electromagnetic integral equations as early as 1972 [5]. The benefits of employing curved-cell models include the fact that often the design and manufacturing process employs such models; furthermore the improvement in accuracy can be substantial [6].

2.1 DIFFERENTIAL GEOMETRY

In all the examples under consideration, the surface of the structure being analyzed is represented by a collection of curved (if necessary) cells or patches. Each patch in x–y–z space is defined by a mapping from a square or triangular reference cell in u–v space. The mapping, and the resulting cell shape, is therefore determined by three functions $x(u, v)$, $y(u, v)$, and $z(u, v)$. These functions are different for each cell of the model. Various ways of systematically obtaining these functions from the surface coordinates of the desired cell are described in the sections of this chapter that follow. One way of describing the curved patch that results from this mapping is by the position vector

$$\bar{r}(u,\ v) = x(u,\ v)\hat{x} + y(u,\ v)\hat{y} + z(u,\ v)\hat{z} \tag{2.1}$$

The vector in (2.1) is directed from the origin $(0, 0, 0)$ to various points (x, y, z) on the curved cell. These points are defined in turn by the parameters u and v. Tangent vectors along the patch may be expressed as a function of u and v as

$$\frac{\partial \bar{r}}{\partial u} = \frac{\partial x}{\partial u}\hat{x} + \frac{\partial y}{\partial u}\hat{y} + \frac{\partial z}{\partial u}\hat{z} \tag{2.2}$$

$$\frac{\partial \bar{r}}{\partial v} = \frac{\partial x}{\partial v}\hat{x} + \frac{\partial y}{\partial v}\hat{y} + \frac{\partial z}{\partial v}\hat{z} \tag{2.3}$$

A normal vector to the patch is obtained from

$$\frac{\partial \bar{r}}{\partial u} \times \frac{\partial \bar{r}}{\partial v} = \left(\frac{\partial y}{\partial u}\frac{\partial z}{\partial v} - \frac{\partial z}{\partial u}\frac{\partial y}{\partial v} \right)\hat{x}$$
$$+ \left(\frac{\partial z}{\partial u}\frac{\partial x}{\partial v} - \frac{\partial x}{\partial u}\frac{\partial z}{\partial v} \right)\hat{y} + \left(\frac{\partial x}{\partial u}\frac{\partial y}{\partial v} - \frac{\partial y}{\partial u}\frac{\partial x}{\partial v} \right)\hat{z} \tag{2.4}$$

If u or v is held constant, the preceding equations describe tangent curves on the surface. Appropriate differential length tangent vectors along these curves are given in x–y–z space by

$$d\bar{s} = \frac{\partial \bar{r}}{\partial u}du \tag{2.5}$$

$$d\bar{t} = \frac{\partial \bar{r}}{\partial v}dv \tag{2.6}$$

The differential surface area is

$$dS = |d\bar{s} \times d\bar{t}| = \left| \frac{\partial \bar{r}}{\partial u} \times \frac{\partial \bar{r}}{\partial v} \right| du\, dv \tag{2.7}$$

The magnitude of the cross product in (2.7) plays the same role as the determinant of a Jacobian matrix, to be defined in Chapter 5. We will denote this quantity by $D(u, v)$. Using (2.4), we obtain

$$D(u, v) = \left| \frac{\partial \bar{r}}{\partial u} \times \frac{\partial \bar{r}}{\partial v} \right|$$

$$= \sqrt{\left(\frac{\partial y}{\partial u} \frac{\partial z}{\partial v} - \frac{\partial z}{\partial u} \frac{\partial y}{\partial v} \right)^2 + \left(\frac{\partial z}{\partial u} \frac{\partial x}{\partial v} - \frac{\partial x}{\partial u} \frac{\partial z}{\partial v} \right)^2 + \left(\frac{\partial x}{\partial u} \frac{\partial y}{\partial v} - \frac{\partial y}{\partial u} \frac{\partial x}{\partial v} \right)^2}$$

$$(2.8)$$

This quantity may be used to express an integral over a curvilinear patch surface in terms of an integral over the reference cell. Suppose that the reference cell occupies the domain $-1 < u < 1$, $-1 < v < 1$. The integral over the curved patch is given by

$$\iint f[x, y, z] dS = \int_{-1}^{1} \int_{-1}^{1} f[x(u, v), \ y(u, v), \ z(u, v)] D(u, v) du \, dv \qquad (2.9)$$

Other uses of $D(u, v)$ in the mapping process will be explored in Chapter 5.

For the electromagnetic analysis of interest, we desire to preserve the first-order vector continuity of the basis functions used within the numerical solution process (either normal or tangential components of the vector functions). In order to accomplish this using the methods of Chapter 5, the surface-patch models must be *conforming*. Conforming cells have their edges and corners aligned so that they yield a continuous surface (although not necessarily a smooth one). If the cells are defined by interpolation polynomials, for instance, adjacent cells must share all the nodes used on the common boundary to define either cell's shape. The corner of one cell will be aligned with the corner of the adjacent cell, not with the center of an adjacent cell's edge, and so on. Nonconforming models might have gaps between adjacent cells, and be unsuitable for providing a continuous expansion.

2.2   MAPPING FROM SQUARE CELLS USING LAGRANGIAN INTERPOLATION POLYNOMIALS

Lagrangian interpolation polynomials find widespread use in computational analysis. A set of M polynomials, each of degree $M - 1$, may be defined on an interval $u_1 < u < u_M$ by the expression

$$P_i^M(u) = \frac{(u - u_1)(u - u_2)\cdots(u - u_{i-1})(u - u_{i+1})\cdots(u - u_M)}{(u_i - u_1)(u_i - u_2)\cdots(u_i - u_{i-1})(u_i - u_{i+1})\cdots(u_i - u_M)} \tag{2.10}$$

where $\{u_i\}$ denote the locations of the interpolation points or *nodes*. These functions satisfy the conditions that

$$P_i^M(u_i) = 1 \tag{2.11}$$

and

$$P_i^M(u_j) = 0, \quad j \neq i \tag{2.12}$$

Since only one function is nonzero at each node, they provide an interpolation. The entire set of M polynomials must be used over each interval to provide a degree $M - 1$ representation.

For example, the two Lagrangian polynomials of degree 1 are given by

$$P_1^2(u) = \frac{(u - u_2)}{(u_1 - u_2)} \tag{2.13}$$

$$P_2^2(u) = \frac{(u - u_1)}{(u_2 - u_1)} \tag{2.14}$$

while the three quadratic polynomials are given by

$$P_1^3(u) = \frac{(u - u_2)(u - u_3)}{(u_1 - u_2)(u_1 - u_3)} \tag{2.15}$$

$$P_2^3(u) = \frac{(u - u_1)(u - u_3)}{(u_2 - u_1)(u_2 - u_3)} \tag{2.16}$$

$$P_3^3(u) = \frac{(u - u_1)(u - u_2)}{(u_3 - u_1)(u_3 - u_2)} \tag{2.17}$$

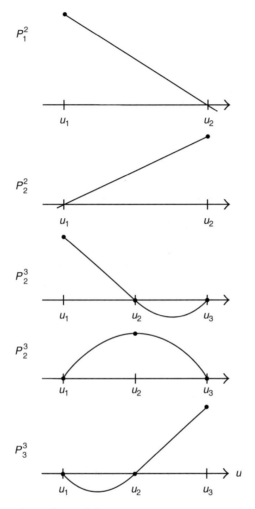

FIGURE 2.1: Lagrangian polynomials.

These are sketched in Fig. 2.1. Although the nodes may be located arbitrarily, in practice they are typically distributed uniformly throughout the interval $u_1 < u < u_M$.

Consider a square reference cell on the domain $-1 < u < 1, -1 < v < 1$, as shown in Fig. 2.2. [The domain $0 < u < 1, 0 < v < 1$ would work equally well; our choice appears more frequently in the finite element literature and will be used throughout this text as the reference coordinates for square domains.] A set of

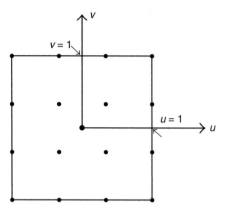

FIGURE 2.2: Interpolation points for $M = 4$.

two-dimensional Lagrangian interpolation polynomials of degree $M - 1$ may be defined by the product rule

$$S_{ij}(u, v) = P_i^M(u) P_j^M(v), \quad i, j = 1, 2, \ldots, M \qquad (2.18)$$

Each of these two-dimensional functions involves M^2 nodes, and each interpolate at one of the nodes. The nodes for $M = 4$ are illustrated in Fig. 2.2 for the case when they are uniformly distributed throughout the domain.

To illustrate the use of Lagrangian interpolation polynomials for the definition of a curvilinear patch, the following section considers the specific case of quadratic polynomials in more detail.

2.3   A SPECIFIC EXAMPLE: QUADRATIC POLYNOMIALS MAPPED FROM A SQUARE REFERENCE CELL

The use of interpolation polynomials facilitates a representation of the desired object in terms of some number of coordinates on its surface. As an example, suppose a piecewise-quadratic model is to be used. It is convenient to define a quadratic cell that coincides with a portion of the actual surface at nine coordinate values. Using the set of Lagrangian interpolation polynomials for $M = 3$, defined by (2.18), a curved patch can be expressed by the coordinates

$$x = \sum_{i=1}^{3} \sum_{j=1}^{3} x_{ij} S_{ij}(u, v)$$

$$y = \sum_{i=1}^{3} \sum_{j=1}^{3} y_{ij} S_{ij}(u, v) \tag{2.19}$$

$$z = \sum_{i=1}^{3} \sum_{j=1}^{3} z_{ij} S_{ij}(u, v)$$

where the functions $\{S_{ij}\}$ denote quadratic Lagrangian interpolation polynomials for $M = 3$. The nine specific interpolation polynomials, assuming that their nodes are distributed uniformly throughout the domain $-1 < u < 1$, $-1 < v < 1$, are

$$S_{11} = \frac{u(u-1)}{2} \frac{v(v-1)}{2} \tag{2.20}$$

$$S_{12} = \frac{u(u-1)}{2}(1+v)(1-v) \tag{2.21}$$

$$S_{13} = \frac{u(u-1)}{2} \frac{v(v+1)}{2} \tag{2.22}$$

$$S_{21} = (1+u)(1-u)\frac{v(v-1)}{2} \tag{2.23}$$

$$S_{22} = (1+u)(1-u)(1+v)(1-v) \tag{2.24}$$

$$S_{23} = (1+u)(1-u)\frac{v(v+1)}{2} \tag{2.25}$$

$$S_{31} = \frac{u(u+1)}{2} \frac{v(v-1)}{2} \tag{2.26}$$

$$S_{32} = \frac{u(u+1)}{2}(1+v)(1-v) \tag{2.27}$$

$$S_{33} = \frac{u(u+1)}{2} \frac{v(v+1)}{2} \tag{2.28}$$

In (2.19), each two-dimensional interpolation polynomial is scaled by a coordinate (x_{ij}, for instance) that usually is an actual value from the surface being modeled. Because of the interpolatory nature of the polynomials, the value of x returned at that node will be that same value (x_{ij}). Thus, the representation in (2.19) will provide an approximation of the surface that coincides with the true surface at the nine nodes. Eight of the nine nodes are located on the cell boundary, with the other node placed in the interior.

Equation (2.19) defines a single cell. Normally, a large number of cells are required to adequately represent a complicated surface. Cells must be smaller in regions where the surface is highly curved to provide an accurate representation.

Each variable within the representation may be explicitly written in terms of the interpolation polynomials. By expanding (2.19), $x(u, v)$ may be written as

$$
\begin{aligned}
x = x_{22} &+ \frac{u}{2}(x_{32} - x_{12}) + \frac{v}{2}(x_{23} - x_{21}) + \frac{uv}{4}(x_{11} - x_{13} - x_{31} + x_{33}) \\
&+ \frac{u^2}{2}(x_{12} - 2x_{22} + x_{32}) + \frac{v^2}{2}(x_{21} - 2x_{22} + x_{23}) \\
&+ \frac{uv^2}{4}(-x_{11} + 2x_{12} - x_{13} + x_{31} - 2x_{32} + x_{33}) \\
&+ \frac{u^2 v}{4}(-x_{11} + 2x_{21} - x_{31} + x_{13} - 2x_{23} + x_{33}) \\
&+ \frac{u^2 v^2}{4}(x_{11} - 2x_{12} + x_{13} - 2x_{21} + 4x_{22} - 2x_{23} + x_{31} - 2x_{32} + x_{33})
\end{aligned}
\tag{2.29}
$$

Derivatives with respect to u and v may be obtained as

$$
\begin{aligned}
\frac{\partial x}{\partial u} = \frac{x_{32} - x_{12}}{2} &+ u(x_{12} - 2x_{22} + x_{32}) + \frac{v}{4}(x_{11} - x_{13} - x_{31} + x_{33}) \\
&+ \frac{uv}{2}(-x_{11} + 2x_{21} - x_{31} + x_{13} - 2x_{23} + x_{33}) \\
&+ \frac{v^2}{4}(-x_{11} + 2x_{12} - x_{13} + x_{31} - 2x_{32} + x_{33}) \\
&+ \frac{uv^2}{2}(x_{11} - 2x_{12} + x_{13} - 2x_{21} + 4x_{22} - 2x_{23} + x_{31} - 2x_{32} + x_{33})
\end{aligned}
\tag{2.30}
$$

and

$$
\begin{aligned}
\frac{\partial x}{\partial v} = \frac{x_{23} - x_{21}}{2} &+ \frac{u}{4}(x_{11} - x_{13} - x_{31} + x_{33}) + v(x_{21} - 2x_{22} + x_{23}) \\
&+ \frac{uv}{2}(-x_{11} + 2x_{12} - x_{13} + x_{31} - 2x_{32} + x_{33}) \\
&+ \frac{u^2}{4}(-x_{11} + 2x_{21} - x_{31} + x_{13} - 2x_{23} + x_{33}) \\
&+ \frac{u^2 v}{2}(x_{11} - 2x_{12} + x_{13} - 2x_{21} + 4x_{22} - 2x_{23} + x_{31} - 2x_{32} + x_{33})
\end{aligned}
\tag{2.31}
$$

Expressions for $y(u, v)$ and $z(u, v)$ and their derivatives are identical to the above, with x replaced by y or z in all occurrences within (2.29)–(2.31), and the subscripts kept the same.

Equations (2.30), (2.31), and the analogous equations involving $y(u, v)$ and $z(u, v)$, provide explicit components of the tangent vectors in (2.2) and (2.3), and the function $D(u, v)$ used in integrals such as (2.9). These quantities are all determined by the mapping of (2.19).

2.4   MAPPING FROM TRIANGULAR CELLS VIA INTERPOLATION POLYNOMIALS

Triangular cells provide additional flexibility when modeling arbitrary surfaces, and also permit a mapping based on interpolation polynomials. Consider the unit right triangle depicted in Fig. 2.3, occupying the domain $0 \leq u \leq 1, 0 \leq v \leq 1$, with $u + v \leq 1$. For this domain, the usual coordinates (u, v) happen to coincide with two of the three so-called *simplex* coordinates (u, v, w) often used to represent quantities on triangles [1–4]. Each simplex coordinate is the relative distance from one side of a triangle to the opposing corner, with value 0 at the side and value 1 at the corner. The third coordinate can be obtained in this case as

$$w = 1 - u - v \tag{2.32}$$

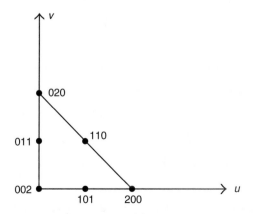

FIGURE 2.3: The indices ijk correspond to the interpolation points of S_{ijk} given in (2.39)–(2.44).

and is clearly linearly dependent on the other two. Simplex coordinates offer a symmetrical way of describing a point within the triangle, and are convenient variables for use in developing interpolation polynomials for triangular domains.

As in the previous sections, interpolation polynomials interpolate at specific nodes within the triangle. For linear polynomials, the obvious interpolation points are the three corners. For a polynomial of degree M, uniformly spaced node locations can be succinctly expressed in terms of simplex coordinates as

$$\left(\frac{i}{M}, \frac{j}{M}, \frac{k}{M} \right), \quad i, j, k = 0, 1, \ldots, M \tag{2.33}$$

with the restriction that $i + j + k = M$. There are a total of

$$\frac{(M+1)(M+2)}{2} \tag{2.34}$$

nodes involved.[1]

Two-dimensional interpolation polynomials are defined in simplex coordinates by the product rule

$$S_{ijk}(u, v, w) = R_i^{(M)}(u) R_j^{(M)}(v) R_k^{(M)}(w) \tag{2.35}$$

where $R_m^{(M)}$ is the Silvester polynomial [3]

$$R_m^{(M)}(u) = \begin{cases} 1 & m = 0 \\ \dfrac{1}{m!} \displaystyle\prod_{i=0}^{m-1}(Mu - i) & m > 0 \end{cases} \tag{2.36}$$

The Silvester polynomial exhibits m equally spaced zeros located at $u = 0, 1/M, \ldots, (m-1)/M$ and has unity value at $u = m/M$. The interpolation polynomial S_{ijk} has unity value at node ijk and vanishes at all the other nodes within the triangle.

Equations (2.35) and (2.36) provide a general framework for interpolation polynomials of any degree. The following section illustrates a specific example.

---

[1] Readers may observe that these nodes are the locations of sample points for Newton–Cotes quadrature rules for triangular domains.

2.5 EXAMPLE: QUADRATIC POLYNOMIALS MAPPED FROM A TRIANGULAR REFERENCE CELL

Consider a curvilinear patch defined in terms of six quadratic interpolation polynomials on a triangular domain, distributed around the reference cell as illustrated in Fig. 2.3. The patch coordinates may be expressed as

$$
\begin{aligned}
x &= \sum_{i=0}^{2} \sum_{j=0}^{2} x_{ijk} S_{ijk}(u, v, w) \\
y &= \sum_{i=0}^{2} \sum_{j=0}^{2} y_{ijk} S_{ijk}(u, v, w) \\
z &= \sum_{i=0}^{2} \sum_{j=0}^{2} z_{ijk} S_{ijk}(u, v, w)
\end{aligned}
\tag{2.37}
$$

where the index k is defined by

$$
k = 2 - i - j \tag{2.38}
$$

As in section 2.3, the coefficients x_{ijk}, y_{ijk}, and z_{ijk} are usually points on the desired surface. From Fig. 2.3, it should be apparent that these points define the boundary of the patch. (For polynomials of greater degree than quadratic, there are also interior points involved.) The six interpolation polynomials in (2.37) are defined

$$
\begin{aligned}
S_{200} &= (2u - 1)u & (2.39) \\
S_{020} &= (2v - 1)v & (2.40) \\
S_{002} &= (2w - 1)w & (2.41) \\
S_{110} &= 4uv & (2.42) \\
S_{101} &= 4uw & (2.43) \\
S_{011} &= 4vw & (2.44)
\end{aligned}
$$

where (u, v, w) denote simplex coordinates. Since the dummy variable w is dependent on the other two, (2.32) may be used to redefine these functions in terms of u and v.

The explicit representation for S_{ijk} may be used to obtain

$$x = x_{002} + u(4x_{101} - 3x_{002} - x_{200}) + v(4x_{011} - 3x_{002} - x_{020})$$
$$+u^2(2x_{200} + 2x_{002} - 4x_{101}) + uv(4x_{002} + 4x_{110} - 4x_{101} - 4x_{011}) \quad (2.45)$$
$$+v^2(2x_{020} + 2x_{002} - 4x_{011})$$

where $y(u, v)$ and $z(u, v)$ have an identical form with x replaced by y or z throughout. Derivatives are obtained as

$$\frac{\partial x}{\partial u} = 4x_{101} - 3x_{002} - x_{200} + u(4x_{200} + 4x_{002} - 8x_{101}) \quad (2.46)$$
$$+v(4x_{002} + 4x_{110} - 4x_{101} - 4x_{011})$$
$$\frac{\partial x}{\partial v} = 4x_{011} - 3x_{002} - x_{020} + u(4x_{002} + 4x_{110} - 4x_{101} - 4x_{011}) \quad (2.47)$$
$$+v(4x_{020} + 4x_{002} - 8x_{011})$$

Derivatives of y and z are similar, with x replaced by y or z on the right-hand sides of (2.46) and (2.47).

Equations (2.46) and (2.47), and the corresponding equations obtained from $y(u, v)$ and $z(u, v)$, provide components of the tangent vectors in (2.2) and (2.3), and the function $D(u, v)$ used in the integral (2.9).

It should be noted that the quadratic interpolation functions defined on quadrilateral cells (Section 2.3) and triangular cells (Section 2.5) exhibit the same behavior along the cell boundaries and thus can be mixed into the same surface model. This combination will be employed with the examples used for illustration in Chapters 6 and 7.

2.6   CONSTRAINTS ON NODE DISTRIBUTION

There are obvious limitations associated with the type of mappings defined in (2.19) and (2.37). For polynomial degrees greater than one, the nodes must be fairly uniformly spaced on the patch in x–y–z space, or the mapping may not be one-to-one: cells may be folded back over themselves. (References [1–4] address this issue.) In the following chapters, we will assume that the node distribution is adequate to ensure a one-to-one mapping. This is straightforward to ensure for simple surface shapes, but may require additional effort for complex bodies.

2.7   HERMITIAN MAPPING FROM SQUARE CELLS

A second limitation associated with the patch-by-patch Lagrangian mappings is that they provide a smooth (highly differentiable) representation within each patch, but do nothing to ensure smoothness beyond first-order continuity across patch boundaries. Electromagnetic fields are sensitive to the presence of corners or edges, and even slight curvature discontinuities. Therefore it may be necessary to consider a means for ensuring patch-to-patch derivative continuity. One possible way of achieving this is to employ a mapping based on Hermite polynomials. We briefly review the procedure for quadrilateral cell shapes.

The lowest degree Hermitian polynomials on the interval $-1 < u < 1$ consist of the four functions

$$H_1^0(u) = \frac{(1-u)^2(2+u)}{4} \tag{2.48}$$

$$H_2^0(u) = \frac{(1+u)^2(2-u)}{4} \tag{2.49}$$

$$H_1^1(u) = \frac{(1-u)^2(1-u)}{4} \tag{2.50}$$

$$H_2^1(u) = \frac{(-1+u)^2(1+u)}{4} \tag{2.51}$$

The functions in (2.48) and (2.49) interpolate at $u = -1$ and $u = +1$, respectively, while those in (2.50) and (2.51) interpolate to the first derivative at those points. (In other words, they equal zero at those points but their derivative equals unity there, while the first derivatives of (2.48) and (2.49) are zero there.) If extended to a square domain $-1 < u < 1$, $-1 < v < 1$, the corresponding two-dimensional functions can be written as

$$S_{ij}^{pq}(u, v) = H_i^p(u)H_j^q(v) \tag{2.52}$$

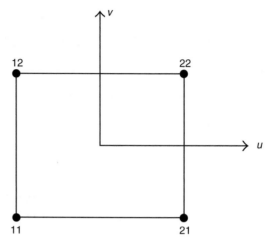

FIGURE 2.4: The indices ij correspond to (2.52).

By allowing indices i and j to equal 1 or 2, while p and q assume values 0 or 1, Eq. (2.52) defines 16 cubic polynomials. Figure 2.4 illustrates the interpolation points in the reference cell.

Using the 16 functions in (2.52) a mapping can be constructed that ensures the continuity of the surface from patch to patch, and its first derivative from patch to patch. Such a mapping has an x-coordinate of the form

$$
x = x_{11} S_{11}^{00}(u, v) + x_{12} S_{12}^{00}(u, v) + x_{21} S_{21}^{00}(u, v) + x_{22} S_{22}^{00}(u, v)
$$

$$
+ \left.\frac{\partial x}{\partial s}\right|_{11} S_{11}^{10}(u, v) + \left.\frac{\partial x}{\partial s}\right|_{12} S_{12}^{10}(u, v) + \left.\frac{\partial x}{\partial s}\right|_{21} S_{21}^{10}(u, v) + \left.\frac{\partial x}{\partial s}\right|_{22} S_{22}^{10}(u, v)
$$

$$
+ \left.\frac{\partial x}{\partial t}\right|_{11} S_{11}^{01}(u, v) + \left.\frac{\partial x}{\partial t}\right|_{12} S_{12}^{01}(u, v) + \left.\frac{\partial x}{\partial t}\right|_{21} S_{21}^{01}(u, v) + \left.\frac{\partial x}{\partial t}\right|_{22} S_{22}^{01}(u, v)
$$

$$
+ \left.\frac{\partial^2 x}{\partial s\, \partial t}\right|_{11} S_{11}^{11}(u, v) + \left.\frac{\partial^2 x}{\partial s\, \partial t}\right|_{12} S_{12}^{11}(u, v) + \left.\frac{\partial^2 x}{\partial s\, \partial t}\right|_{21} S_{21}^{11}(u, v) + \left.\frac{\partial^2 x}{\partial s\, \partial t}\right|_{22} S_{22}^{11}(u, v)
$$

$$
\tag{2.53}
$$

where s and t are tangential variables (with units of length) along the directions of parameters u and v. (These may be replaced by u and v, or any other convenient parameters, as long as the interpretation of the coefficients of the derivative terms is not an issue.) Similar expressions can be constructed for y and z. Explicit derivatives must be obtained from (2.53), following an approach similar to that used in Sections

2.3 and 2.5, to obtain $D(u, v)$ in (2.9) and other parameters to be identified in Chapter 5.

2.8 CONNECTIVITY

The representation of a surface in terms of cells involves more than just a list of the coordinates (nodes). The model also requires *connectivity*. In its simplest form, connectivity is provided by a pointer array that identifies the specific nodes belonging to each cell (Fig. 2.5). Each node and each cell in the model are assigned an integer; the connectivity array assigns a group of nodes to each cell. In practice, the connectivity array simultaneously provides an ordering of the nodes within each cell to eliminate ambiguity as to the cell orientation, and to clearly indicate which

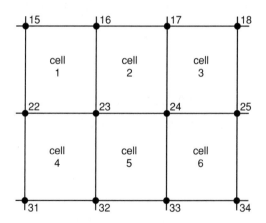

cell index	node 1	node 2	node 3	node 4
1	22	23	15	16
2	23	24	16	17
3	24	25	17	18
4	31	32	22	23
5	32	33	23	24
6	33	34	24	25
.				
.				
.				

FIGURE 2.5: Portion of mesh and pointer indicating the cell-to-node connectivity.

face of the cell is inside or outside. For cells with relatively few nodes, a clockwise or counter-clockwise orientation of nodes around the cell center may be adequate for this purpose. For scalar problems (and especially in connection with scalar finite element solutions of differential equations), the cell-to-node connectivity is sufficient to describe the geometry of the structure. It is often convenient to incorporate a separate list that identifies nodes residing on the outer boundary of the domain of interest.

For problems involving vector quantities, it is usually necessary to extend the connectivity to the cell edges. Each edge in the model is assigned an integer index. Additional pointers may be generated that identify edges associated with cells, nodes associated with edges, or cells associated with edges. For simple domains, these pointers can be generated automatically from the basic cell-to-node pointer described above. Their availability simplifies the computer program that performs the analysis.

REFERENCES

[1]   D. S. Burnett, *Finite Element Analysis*. Reading, MA: Addison-Wesley, 1987.

[2]   O. C. Zienkiewicz and R. L. Taylor, *The Finite Element Method*. London: McGraw-Hill, 1988.

[3]   P. P. Silvester and R. L. Ferrari, *Finite Elements for Electrical Engineers*. Cambridge: Cambridge University Press, 1990.

[4]   J. E. Akin, *Finite Elements for Analysis and Design*. San Diego, CA: Academic Press, 1994.

[5]   D. L. Knepp and J. Goldhirsh, "Numerical analysis of electromagnetic radiation properties of smooth conducting bodies of arbitrary shape," *IEEE Trans. Antennas Propagat.*, vol. AP-20, pp. 383–388, 1972. doi:10.1109/TAP.1972.1140210

[6]   M. I. Sancer, R. L. McClary, and K. J. Glover, "Electromagnetic computation using parametric geometry" *Electromagnetics*, vol. 10, pp. 85–103, 1990.

CHAPTER 3

Divergence-Conforming
Basis Functions

The vector basis functions used in electromagnetic modeling usually belong in one of two classes: divergence-conforming and curl-conforming. This chapter considers the former, as defined for square and triangular reference cells. Curl-conforming functions will be the topic of Chapter 4. As will be explained in Chapter 4, these two types of functions are closely related, and therefore developments and extensions of basis functions of one type motivate a simultaneous development for the other type. Similarly, the literature on these two types of basis functions includes many interrelated articles.

Most integral equation numerical techniques reported during the past two decades have employed $p = 0$ divergence conforming basis functions. The $p = 0$ functions for square and triangular cells are presented below, and used for illustration in the EFIE implementation of Chapter 6. However, the advantages associated with curvilinear cell shapes are not likely to be fully realized without the use of higher order bases. Therefore, methods for constructing basis functions of higher polynomial degrees are also described and tables of these functions are included.

3.1   CHARACTERISTICS OF VECTOR FIELDS
AND VECTOR BASIS FUNCTIONS

The term *field* is used in engineering to denote any function of position (which may also be a function of time). A *scalar field* is a function associating a parameter or number (some scalar quantity) with each point. One example of a scalar field is the temperature throughout some region. Electromagnetic fields are *vector* quantities, meaning that a direction in $x-y-z$ space is also associated with their value at each point. The velocity of water throughout a river is another example of a vector field. In the present context the vector fields are phasor quantities representing a sinusoidal time dependence, and are consequently complex valued functions of position, with a direction at each point.

Vector fields exhibit two properties: divergence and curl. *Divergence* refers to the tendency of the field to spread apart at any point, and is obtained for a general 3D vector $\bar{B}$ by the operation

$$\nabla \cdot \bar{B} = \frac{\partial B_x}{\partial x} + \frac{\partial B_y}{\partial y} + \frac{\partial B_z}{\partial z} \tag{3.1}$$

It is also of interest to consider the surface divergence of a vector function $\bar{B}$ that is tangential to some surface, which can be expressed as

$$\nabla_s \cdot \bar{B} = \frac{\partial B_s}{\partial s} + \frac{\partial B_t}{\partial t} \tag{3.2}$$

where s and t are local orthogonal variables with units of length along the surface. The divergence is a scalar quantity.

The *curl* of a vector field is related to the tendency of the field to rotate or twist about a point, and will be defined mathematically in Chapter 4. Since there are three planes in which the vector field can twist, the curl of a general 3D vector is itself expressed as a vector quantity. It is also possible to define a surface curl operation, which can be thought of as a scalar property associated with a vector function tangential to the surface.

The importance of the divergence and curl of electromagnetic fields should be apparent from a study of the key equations describing electromagnetic fields—those named for James Clerk Maxwell. In their modern form for phasor fields, they relate the curl of each field to the other field and to a current density source, and the divergence of each field to a charge density source.

Within numerical techniques for electromagnetics, basis functions are used to represent the fields directly or the surface currents that produce the fields. Historically, numerical procedures of this type were initially developed for scalar quantities, such as temperature in heat transfer problems or the voltage field in electrostatics. It was natural to develop families of scalar basis functions for the purpose of representing scalar fields. In principle, each component of a vector field can also be represented by scalar basis functions. However, there are difficulties with representing vector fields this way. For general cell shapes, a component-by-component representation is not convenient for imposing continuity and boundary conditions. Nor does such a representation allow the user to easily control the divergence or curl of the quantities being represented.

In recent years, a wide variety of vector basis functions have been developed to address some of the apparent drawbacks of scalar representations of vector fields. Each basis function provides a local vector direction typically oriented along or across cell edges. The simplest functions exhibit either a constant divergence or a constant curl within a cell. The basis functions fall into two classes: divergence-conforming and curl-conforming functions.

3.2   WHAT DOES DIVERGENCE-CONFORMING MEAN?

The term *conforming* was defined in Chapter 2, where it was used to describe a subsectional model of a surface that maintained first order continuity (no gaps between cells). The term *divergence-conforming* is used to denote a function $\bar{B}$ that maintains the first-order continuity needed by the divergence operator in (3.1) or (3.2). If this condition is met, it will result in the complete absence of Dirac delta functions in $\nabla \cdot \bar{B}$. In practice, discontinuities associated with a basis function for subsectional cells occur at the cell edges. The divergence operator involves a differentiation along the vector direction of $\bar{B}$. Since the normal-vector component of the function is differentiated in the normal direction (across the cell edge), the important continuity is that of the normal component. It does not matter if the function maintains cell-to-cell tangential-vector continuity, since the divergence operation does not differentiate the tangential component across the edge. Consequently, *a divergence-conforming basis function is one that maintains first-order normal-vector continuity across cell edges*, but not necessarily any tangential-vector continuity.

The complementary behavior is provided by *curl-conforming* basis functions. As described in Chapter 4, curl-conforming functions maintain tangential-vector continuity across cell edges.

3.3   HISTORY OF THE USE OF DIVERGENCE-CONFORMING BASIS FUNCTIONS

Divergence-conforming basis functions were first investigated for finite element solutions by Raviart and Thomas [1]. The lowest order ($p = 0$) functions were developed independently by Glisson for numerical solutions of the EFIE [2], and they have been widely used within the electromagnetics community [3, 4]. These low-order functions were generalized to arbitrary polynomial degree p by Nedelec [5], initially for use with time domain finite element solutions of Maxwell's equations. Nedelec proposed a family of divergence-conforming spaces that could be adapted to basis functions of any polynomial degree, for cells of triangular, tetrahedral, quadrilateral, or hexahedral shape. The $p = 0$ functions for rectangular and triangular cells are known as "rooftop" (or "triangular rooftop") functions, because of their obvious similarity to the roof of an A-frame house (Fig. 3.1). The triangular cell $p = 0$ functions are also known as RWG basis functions after the authors of [4].

Most types of higher order basis functions are either *interpolatory* or *hierarchical*. The coefficient of each interpolatory function represents one vector component of the unknown quantity at a specific point within the domain. The basis function "interpolates" at that point. Interpolatory representations usually involve a number of basis functions of the same polynomial degree p. To improve the order of the representation, all the interpolatory functions are exchanged for functions of greater

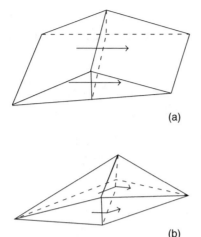

(a)

(b)

FIGURE 3.1: Rooftop function straddling two square cells (a) or two triangular cells (b). The vector direction is roughly indicated by the arrows.

degree. Interpolatory functions offer relatively good linear independence, and their coefficients have the simple physical interpretation as being the unknown current density or field at various locations.

Hierarchical functions, in contrast, are designed to build systematically upon basis functions of lower order. If the lowest-order members provide a constant representation, the next order will provide the linear degrees of freedom, followed by the quadratic, and so on. Each order is added to the previous orders, so the computations associated with the lower order functions may not need to be repeated as the representation is refined. Hierarchical functions make it more efficient to refine the representation to higher order, but usually provide weaker linear independence than interpolatory functions. In addition, the coefficients of hierarchical functions have no physical interpretation (the full set of functions must be superimposed to obtain the field or current density at some location). It is more difficult in some cases to ensure the satisfaction of boundary conditions or continuity conditions with hierarchical representations [6].

A number of different interpolatory and hierarchical families of higher order basis functions have been proposed for quadrilateral and triangular cell shapes. Interpolatory functions of arbitrary order, which have the rooftop type of basis as the $p = 0$ members, will be summarized in the sections following immediately. Hierarchical functions are considered in Sections 3.9 and 3.10.

3.4   BASIS FUNCTIONS OF ORDER $p = 0$ FOR A SQUARE REFERENCE CELL

Figure 3.2 shows a square reference cell occupying the region $-1 < u < 1$, $-1 < v < 1$. Four divergence-conforming basis functions of the lowest order can be defined within this cell as

$$\bar{R}_1^{\text{div}} = \frac{u-1}{2}\hat{u} \tag{3.3}$$

$$\bar{R}_2^{\text{div}} = \frac{u+1}{2}\hat{u} \tag{3.4}$$

$$\bar{R}_3^{\text{div}} = \frac{v-1}{2}\hat{v} \tag{3.5}$$

$$\bar{R}_4^{\text{div}} = \frac{v+1}{2}\hat{v} \tag{3.6}$$

Each of these functions has a nonzero normal-vector component along one cell edge, and contributes no normal component along any other edge. The normal components are constant, and it is convenient to think of each function as interpolating at the center of the appropriate edge. As an example, the function $\bar{R}_1$ is zero

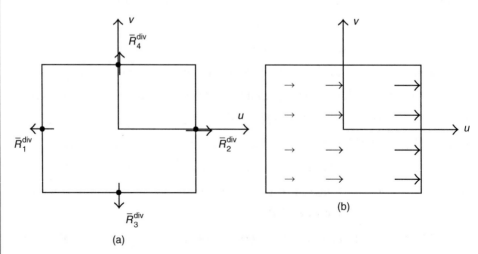

FIGURE 3.2: (a) The four $p = 0$ divergence-conforming basis functions on a square cell. (b) The function $\bar{R}_2^{\text{div}}$.

at $u = +1$, and is entirely tangential at $v = \pm 1$. Its normal component at $u = -1$ is directed out of the cell (in the $-\hat{u}$ direction).

For these functions to be divergence-conforming, they must be matched with basis functions in adjacent cells that maintain the continuity of the nonzero normal component. Conceptually, functions similar to (3.3)–(3.6) are assigned to the adjacent cells for this purpose, and attached to these in order to construct the rooftop functions of Fig. 3.1. The sign of some of the basis functions must be adjusted so that all maintain the same normal-vector direction across cell boundaries. Divergence-conforming functions usually do not attempt to maintain the continuity of the tangential component of the vector unknown, and a rooftop representation will generally exhibit jump discontinuities in the tangential components at cell edges.

The coefficients of the preceding basis functions have an obvious physical interpretation: they are the field or current density component at the center of the cell edge, perpendicular to that edge. For this interpretation to survive the process of mapping these functions to curvilinear cells, their normalization in the x–y–z space must be maintained so that they have a unit normal component at the appropriate interpolation point on the cell edge. The proper normalization will be determined in connection with the mapping process discussed in Chapter 5.

The divergence of the preceding functions in the reference cell coordinates may be calculated using

$$\nabla \bullet \bar{R} = \frac{\partial R_u}{\partial u} + \frac{\partial R_v}{\partial v} \tag{3.7}$$

For the functions in (3.3)–(3.6), the reference cell divergence is constant:

$$\frac{\partial R_{1u}^{\text{div}}}{\partial u} + \frac{\partial R_{1v}^{\text{div}}}{\partial v} = \frac{1}{2} \tag{3.8}$$

$$\frac{\partial R_{2u}^{\text{div}}}{\partial u} + \frac{\partial R_{2v}^{\text{div}}}{\partial v} = \frac{1}{2} \tag{3.9}$$

$$\frac{\partial R_{3u}^{\text{div}}}{\partial u} + \frac{\partial R_{3v}^{\text{div}}}{\partial v} = \frac{1}{2} \tag{3.10}$$

$$\frac{\partial R_{4u}^{\text{div}}}{\partial u} + \frac{\partial R_{4v}^{\text{div}}}{\partial v} = \frac{1}{2} \tag{3.11}$$

After being mapped to a curvilinear cell in x–y–z space, the divergence of the basis functions also incorporates a (nonconstant) scale factor related to the Jacobian of the mapping, a possible adjustment in sign, and a normalization factor. The appropriate divergence calculation is considered in Chapter 5.

3.5   BASIS FUNCTIONS OF ORDER $p = 0$ FOR A TRIANGULAR REFERENCE CELL

Figure 3.3 shows a triangular reference cell occupying the region $0 < u < 1, 0 < v < 1, u + v < 1$. Three divergence conforming basis functions of the lowest order can be defined within this cell as

$$\bar{R}_1^{\text{div}} = (u - 1)\hat{u} + v\hat{v} \tag{3.12}$$

$$\bar{R}_2^{\text{div}} = u\hat{u} + (v - 1)\hat{v} \tag{3.13}$$

$$\bar{R}_3^{\text{div}} = \sqrt{2}(u\hat{u} + v\hat{v}) \tag{3.14}$$

Each of these functions interpolates to the vector component normal to one edge, and contributes no normal component along any other edge. For example, the function $\bar{R}_1$ contributes a constant normal component along $u = 0$ axis, and is

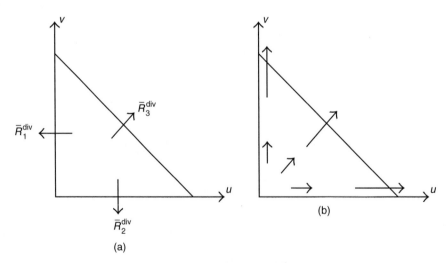

(a)

(b)

FIGURE 3.3: (a) The three $p = 0$ divergence-conforming basis functions on a triangular cell. (b) The function $\bar{R}_3^{\text{div}}$.

entirely tangential or zero along the other two edges. Its normal component at $u = 0$ is directed out of the cell in the $-\hat{u}$ direction. Function $\bar{R}_2$ provides an outward normal component along the $v = 0$ edge, while $\bar{R}_3$ contributes a constant normal component along the edge at $u + v = 1$.

For these functions to be divergence-conforming, they must be matched with basis functions in adjacent cells to create the triangular rooftop functions of Fig. 3.1. In the process, the sign of the local basis function in each cell must be adjusted to maintain a consistent normal-vector direction across cell boundaries. The tangential components of (3.12)–(3.14) are linear functions along cell boundaries; cell-to-cell tangential continuity is not maintained by the triangular rooftop functions.

As in the square cell case, the coefficients of the preceding basis functions have the physical interpretation that they are the normal component of the field or current density at the center of each cell edge. The proper normalization is considered in Chapter 5.

In the local coordinates of the reference cell, these functions have divergence

$$\frac{\partial R_{1u}^{\text{div}}}{\partial u} + \frac{\partial R_{1v}^{\text{div}}}{\partial v} = 2 \tag{3.15}$$

$$\frac{\partial R_{2u}^{\text{div}}}{\partial u} + \frac{\partial R_{2v}^{\text{div}}}{\partial v} = 2 \tag{3.16}$$

$$\frac{\partial R_{3u}^{\text{div}}}{\partial u} + \frac{\partial R_{3v}^{\text{div}}}{\partial v} = 2\sqrt{2} \tag{3.17}$$

The surface divergence on curvilinear cells in x–y–z space is different due to a nonconstant scale factor, a normalization factor, and a possible change in sign (Chapter 5).

3.6 NEDELEC'S MIXED-ORDER SPACES AND THE EFIE

The $p = 0$ basis functions of (3.3)–(3.6) and (3.12)–(3.14) correspond to the lowest order members of Nedelec's mixed-order divergence-conforming spaces [5]. Unlike most representations that are mathematically *complete* to the same polynomial degree in each variable, these spaces deliberately provide a representation with one additional polynomial degree along the primary vector direction of the basis function, compared with the orthogonal direction. [For instance, the $p = 0$ functions provide a linear dependence for the tangential components along cell edges, but only a constant dependence in the normal components.] As a consequence, the divergence of each function is complete to the same degree as the basis function itself. From another point of view, Nedelec's mixed-order spaces discard some of the degrees of freedom from a polynomial-complete expansion. When used to discretize an equation whose leading-order derivative is a divergence operator, the discarded degrees of freedom are essentially those that do not contribute to the balance of terms in the discretized equation. The EFIE is one such equation.

The EFIE of Eq. (1.8) is usually manipulated into a more suitable form for discretization (see Chapter 6). The resulting equation involves two terms: one proportional to the surface current density and the other proportional to the divergence of the surface current density. The latter quantity is essentially the electric charge density, related to the current density by

$$\rho_s = \frac{-1}{j\omega} \nabla_s \bullet \bar{J} \tag{3.18}$$

Because of the nature of the EFIE operator, the charge density provides the dominant contribution to the electric field near the source, while the current density is the dominant contributor to the electric field far from the source. The accuracy of the overall numerical solution is limited by the near-field interactions, or equivalently by the representation for charge density. Although the current is the quantity that is explicitly represented by the basis functions, the behavior of the divergence of those basis functions (the charge density) is critical. In fact, the charge representation appears to be more important than the current representation.

When used to expand the current density $\bar{J}$, in conjunction with (3.18) for ρ_s, Nedelec's mixed-order spaces provide a mathematically complete representation of the surface charge density. With mixed-order basis functions used for $\bar{J}$, the charge associated with the $\hat{s}$-component of $\bar{J}$ and the charge associated with the $\hat{t}$-component of $\bar{J}$ are both complete to the same degree in each variable. This interpretation may explain the longstanding observation that mixed-order solutions of the EFIE are more accurate than those obtained using polynomial-complete expansions for $\bar{J}$ [7, 8].

Nedelec's mixed-order divergence-conforming spaces appear ideally suited for use with the EFIE. Consequently, we restrict our consideration to basis functions of this type.

3.7   HIGHER-ORDER INTERPOLATORY FUNCTIONS FOR SQUARE CELLS

For constructing higher-order basis functions, it will be convenient to introduce the *shifted Silvester polynomial*

$$S_m^{(M)}(\xi) = \begin{cases} 1 & m = 1 \\ \dfrac{1}{(m-1)!} \displaystyle\prod_{i=1}^{m-1}(M\xi - i) & 2 \le m \le M+1 \end{cases} \qquad (3.19)$$

for use on the domain $0 \le \xi \le 1$ [9]. This polynomial is related to the Silvester polynomial of (2.36) through

$$S_m^{(M)}(\xi) = R_{m-1}^{(M)}\left(\xi - \frac{1}{M}\right) \qquad (3.20)$$

The zeros of several shifted Silvester polynomials are plotted in Fig. 3.4. The polynomial in (3.19) is of degree $m - 1$, and divides the domain $0 \le \xi \le 1$ into M intervals. Unlike the Silvester polynomial of (2.36), however, (3.19) has no zero at $\xi = 0$.

Nedelec's prescription for quadrilateral spaces requires $2(p + 1)(p + 2)$ degrees of freedom for a representation of minimum degree p [5]. The $p = 0$ representation of (3.3)–(3.6) involves four functions, conveniently distributed so that each one provides a perpendicular vector component at one edge. Higher order functions associate some of the degrees of freedom with each edge and some with

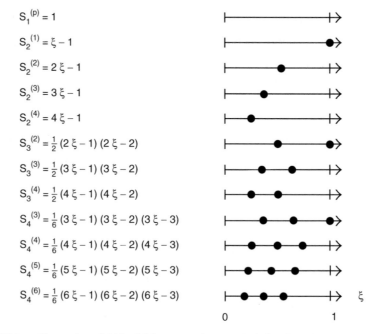

$$S_1^{(p)} = 1$$

$$S_2^{(1)} = \xi - 1$$

$$S_2^{(2)} = 2\,\xi - 1$$

$$S_2^{(3)} = 3\,\xi - 1$$

$$S_2^{(4)} = 4\,\xi - 1$$

$$S_3^{(2)} = \tfrac{1}{2}\,(2\,\xi - 1)\,(2\,\xi - 2)$$

$$S_3^{(3)} = \tfrac{1}{2}\,(3\,\xi - 1)\,(3\,\xi - 2)$$

$$S_3^{(4)} = \tfrac{1}{2}\,(4\,\xi - 1)\,(4\,\xi - 2)$$

$$S_4^{(3)} = \tfrac{1}{6}\,(3\,\xi - 1)\,(3\,\xi - 2)\,(3\,\xi - 3)$$

$$S_4^{(4)} = \tfrac{1}{6}\,(4\,\xi - 1)\,(4\,\xi - 2)\,(4\,\xi - 3)$$

$$S_4^{(5)} = \tfrac{1}{6}\,(5\,\xi - 1)\,(5\,\xi - 2)\,(5\,\xi - 3)$$

$$S_4^{(6)} = \tfrac{1}{6}\,(6\,\xi - 1)\,(6\,\xi - 2)\,(6\,\xi - 3)$$

FIGURE 3.4: Examples of shifted Silvester polynomials defined on the interval $0 \le \xi \le 1$. The figure at right shows the location of the zeros of the polynomial at left.

the cell interior, as summarized in Table 3.1. There are a variety of forms that the corresponding basis functions can assume. For the $p = 1$ case, there are eight functions associated with cell edges, where they provide a linear normal behavior, and four functions that contribute no normal component to the edges. (The $p = 1$ functions collectively provide a quadratic tangential dependence along the cell edges.) Figure 3.5 shows one approach for assigning the interpolation points for the 12 $p = 1$ basis functions, after [9]. Specific functions can be obtained by combining the shifted Silvester polynomials from Fig. 3.4 to produce zeros at all the interpolation points except one.

As an example, the $p = 1$ function $\bar{R}_3$ that interpolates at the point $u = 1, v = 1/3$ in Fig. 3.5 can be constructed from the combination

$$\bar{R}_3(u, v) = \hat{u}\, S_3^{(2)}\left(\frac{1 - u}{2}\right) S_2^{(3)}\left(\frac{1 + v}{2}\right) \tag{3.21}$$

TABLE 3.1: Number of Degrees of Freedom and Polynomial Behavior of Normal and Tangential Vector Components Associated with Nedelec's Spaces of Minimal Degree p for Quadrilateral Cells.

p	$\dfrac{2(p+1)}{(p+2)}$	NORMAL	TANGENTIAL	# ON EDGES	# IN CELL
0	4	Constant	Linear	4	0
1	12	Linear	Quadratic	8	4
2	24	Quadratic	Cubic	12	12
3	40	Cubic	Degree 4	16	24
4	60	Degree 4	Degree 5	20	40
5	84	Degree 5	Degree 6	24	60

Note. The column labeled "# on edges" indicates the number of basis functions associated with the normal component at the cell edges, which must be made normally continuous with analogous functions in adjacent cells. The remaining degrees of freedom ("# in cell") are entirely local and are not tied to functions in adjacent cells.

where the form of the arguments adapts the domain $0 \leq \xi \leq 1$ to either $-1 < u < 1$ or $-1 < v < 1$. The shifted Silvester polynomial $S_3^{(2)}$ equals zero when its argument is $1/2$ or 1, meaning that (3.21) is zero at $u = -1$ and $u = 0$. The polynomial $S_2^{(3)}$ assigns a zero when its argument is $1/3$, or when $v = -1/3$. Thus, the combination in (3.21) places zeros at the five other interpolation points indicated in Fig. 3.5. At the interpolation point $u = 1$, $v = 1/3$, (3.21) has unit magnitude. The basis function is of degree 2 in u and degree 1 in v.

As a second example, the $p = 1$ function $\bar{R}_5$, interpolating at $u = 0$, $v = -1/3$ in Fig. 3.5 can be constructed as

$$\bar{R}_5(u, v) = \hat{u} S_2^{(1)}\left(\frac{1-u}{2}\right) S_2^{(1)}\left(\frac{1+u}{2}\right) S_2^{(3)}\left(\frac{1-v}{2}\right) \tag{3.22}$$

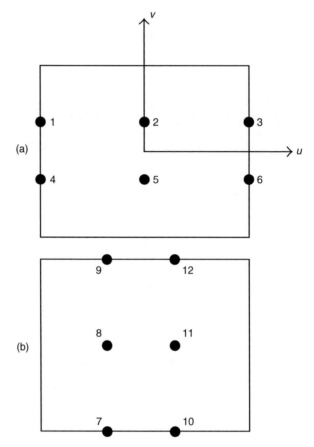

FIGURE 3.5: Interpolation points for the 12 $p = 1$ basis functions. (a) Points for $\hat{u}$ components, (b) points for $\hat{v}$ components.

The polynomial $S_2^{(1)}$ is zero when its argument equals 1, therefore the function

$$S_2^{(1)}\left(\frac{1-u}{2}\right) S_2^{(1)}\left(\frac{1+u}{2}\right) \tag{3.23}$$

places zeros at $u = -1$ and $u = +1$. The polynomial $S_2^{(3)}$ assigns a zero when its argument is $1/3$, or when $v = 1/3$ in (3.22). Thus, the construction in (3.22) places zeros at all the interpolation points except the point labeled "5" in Fig. 3.5. The function $\bar{R}_5$ is of degree 2 in u and degree 1 in v, and has a value of $1/4$ at the interpolation point. Therefore, it must be scaled by a factor of 4 in order to exhibit a unit value there.

Using the preceding approach, the set of $p = 1$ basis functions consists of

$$\bar{R}_1 = (-\hat{u})\frac{u(u-1)(3v+1)}{4} \tag{3.24}$$

$$\bar{R}_2 = \hat{u}\frac{(1-u)(1+u)(3v+1)}{2} \tag{3.25}$$

$$\bar{R}_3 = \hat{u}\frac{u(u+1)(3v+1)}{4} \tag{3.26}$$

$$\bar{R}_4 = (-\hat{u})\frac{u(u-1)(1-3v)}{4} \tag{3.27}$$

$$\bar{R}_5 = \hat{u}\frac{(1-u)(1+u)(1-3v)}{2} \tag{3.28}$$

$$\bar{R}_6 = \hat{u}\frac{u(u+1)(1-3v)}{4} \tag{3.29}$$

$$\bar{R}_7 = (-\hat{v})\frac{(1-3u)v(v-1)}{4} \tag{3.30}$$

$$\bar{R}_8 = \hat{v}\frac{(1-3u)(1-v)(1+v)}{2} \tag{3.31}$$

$$\bar{R}_9 = \hat{v}\frac{(1-3u)v(v+1)}{4} \tag{3.32}$$

$$\bar{R}_{10} = (-\hat{v})\frac{(3u+1)v(v-1)}{4} \tag{3.33}$$

$$\bar{R}_{11} = \hat{v}\frac{(3u+1)(1-v)(1+v)}{2} \tag{3.34}$$

$$\bar{R}_{12} = \hat{v}\frac{(3u+1)v(v+1)}{4} \tag{3.35}$$

These functions have been normalized to unity at the interpolation points, and are adjusted so that those with interpolation points on the edges point out of the cell. The edge-based functions must be paired with corresponding functions in the adjacent cells to maintain normal-vector continuity between cells.

The $p = 2$ case involves 24 basis functions per cell, with 12 that are edge-based (and interpolate to normal components on edges) and 12 that contribute no

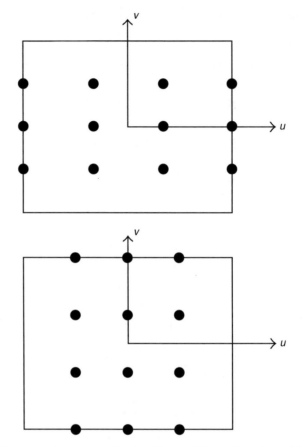

FIGURE 3.6: Interpolation points for the 24 $p = 2$ basis functions. (a) $\hat{u}$ components, (b) $\hat{v}$ components.

normal components along cell edges. (Since the edge-based functions are shared with neighboring cells, the global number of unknowns is less than 24 times the number of cells.) The overall representation is quadratic for the normal vector components, and cubic for the tangential components. Figure 3.6 illustrates one possibility for the 24 interpolation points, based on [9]. The construction of specific basis functions follows in the same manner as the $p = 1$ functions discussed above. Additional information and alternative expressions may be found in [9]. We note that there are other ways of constructing interpolatory vector bases that may offer advantages such as improved matrix condition numbers [10].

3.8   HIGHER-ORDER INTERPOLATORY FUNCTIONS FOR TRIANGULAR CELLS

Interpolatory basis functions of the divergence-conforming type may be developed for triangular cells in much the same manner as those for square cells in Section 3.6. Since there are only three edges per cell, it is possible to eliminate additional degrees of freedom compared to the basis functions for square cells, while ensuring that the divergence of the function is mathematically complete to the same degree as the basis function itself. Nedelec's spaces require a total of $(p + 1)(p + 3)$ degrees of freedom for a representation of minimum degree p [5], with $3(p + 1)$ of those degrees of freedom contributing a nonzero normal component at the cell edges.

The $p = 0$ basis functions are those in (3.12)–(3.14). Each function exhibits a constant normal-vector component on one edge of the triangle, and can be thought of as interpolating to the normal-vector function at the center of the appropriate edge.

Reference [9] proposed a systematic approach for higher-degree representations, as illustrated for the $p = 1$ and $p = 2$ cases in Fig. 3.7. Interpolation points are defined at the cell edges and at regularly spaced intervals within the cell. One basis function is assigned to each of the edge points, while two are assigned at each interior point. These basis functions are constructed from products of the shifted Silvester polynomials with the $p = 0$ functions of (3.12)–(3.14). The $p = 1$ representation involves two basis functions per edge, and two functions that can be made to interpolate at the interior point, for a total of eight degrees of freedom.

For example, consider the $p = 1$ basis function for node 201 in Fig. 3.7(a). This basis function builds on the $p = 0$ function $\bar{R}_2^{\text{div}}$ in (3.13), which provides the underlying vector direction. However, the $p = 1$ function must also vanish at the points labeled 102 and 111 in Figure 3.7(a), meaning that it must be zero at $u = 1/3$. (It is not necessary to zero the function at points 012, 021, 120, or 210 since $\bar{R}_2^{\text{div}}$ has no normal vector component at those locations.) The desired behavior may be obtained from the shifted Silvester polynomial $S_2^{(3)}$. Thus, the

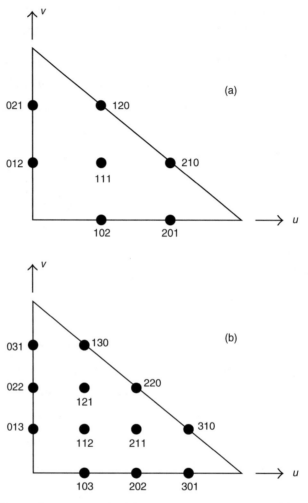

FIGURE 3.7: Interpolation points for (a) $p = 1$ basis functions and (b) $p = 2$ basis functions. Two functions interpolate at each of the interior nodes.

basis function may be constructed as

$$\bar{R}_{201}(u, v) = \bar{R}_2^{\text{div}} S_2^{(3)}(u) \tag{3.36}$$

where the arguments range over $0 < u < 1$ and $0 < v < 1$. It is easily verified that (3.36) has a linear normal component along the edge $v = 0$, and no normal component along the other two edges of the cell. The tangential-vector component is quadratic. This function must be matched with a basis function in the adjacent cell that also has a linear normal component in the same global direction.

Two functions must be assigned to interpolate at the interior node labeled 111 in Fig. 3.7(a). These functions may utilize any two of the three $p = 0$ basis functions in (3.12)–(3.14), multiplied by shifted Silvester polynomials to zero the result at the other interpolation points. There are three suitable constructions possible:

$$\bar{R}_{111a} = \bar{R}_1^{\text{div}} S_2^{(1)}(1 - u) \tag{3.37}$$

$$\bar{R}_{111b} = \bar{R}_2^{\text{div}} S_2^{(1)}(1 - v) \tag{3.38}$$

$$\bar{R}_{111c} = \bar{R}_3^{\text{div}} S_2^{(1)}(u + v) \tag{3.39}$$

However, only two of these are linearly independent functions, so one must be discarded. The functions in (3.37)–(3.39) each have an identically zero normal component at all three of the cell edges.

The set of eight $p = 1$ basis functions obtained from the preceding construction is as follows:

$$\bar{R}_{012} = \{(u - 1)\hat{u} + v\hat{v}\}(3w - 1) \tag{3.40}$$
$$= \{(u - 1)\hat{u} + v\hat{v}\}(2 - 3u - 3v)$$

$$\bar{R}_{021} = \{(u - 1)\hat{u} + v\hat{v}\}(3v - 1) \tag{3.41}$$

$$\bar{R}_{102} = \{u\hat{u} + (v - 1)\hat{v}\}(3w - 1) \tag{3.42}$$
$$= \{u\hat{u} + (v - 1)\hat{v}\}(2 - 3u - 3v)$$

$$\bar{R}_{201} = \{u\hat{u} + (v - 1)\hat{v}\}(3u - 1) \tag{3.43}$$

$$\bar{R}_{120} = \left\{ \sqrt{2}(u\hat{u} + v\hat{v}) \right\}(3v - 1) \tag{3.44}$$

$$\bar{R}_{210} = \left\{ \sqrt{2}(u\hat{u} + v\hat{v}) \right\}(3u - 1) \tag{3.45}$$

$$\bar{R}_{111a} = \{(u - 1)\hat{u} + v\hat{v}\}(-u) \tag{3.46}$$

$$\bar{R}_{111b} = \{u\hat{u} + (v - 1)\hat{v}\}(-v) \tag{3.47}$$

The edge-based functions in (3.40)–(3.45) are normalized so that the outward component is unity at the interpolation point. These functions must be paired with analogous functions in adjacent cells to ensure normal-vector continuity at cell boundaries. The cell-based functions in (3.46) and (3.47) are not normalized, and are not paired with functions outside the cell.

The preceding idea may be extended to develop basis functions of any degree p. Additional information and alternative expressions may be found in [9].

3.9   HIGHER-ORDER HIERARCHICAL FUNCTIONS FOR SQUARE CELLS

Because interpolatory basis functions of one degree are not re-used in representations of greater polynomial degree, hierarchical functions are more computationally efficient when p-refinement is used to obtain a more accurate solution. Hierarchical vector functions can be constructed to satisfy the requirements of Nedelec's mixed-order spaces. It is also desirable to construct functions that provide reasonable linear independence (to avoid ill-conditioned matrices). Ideally, the higher-order functions should not duplicate the principal degrees of freedom provided by the lower-order functions, in order to provide a representation where the coefficients of lower-order functions only change slightly as the higher-order functions are added. (This situation suggests some type of orthogonality between lower-order and higher-order functions.) To date, these constraints are only partially realized in the available basis sets.

Hierarchical vector basis functions for quadrilateral or hexahedral cell shapes have been proposed by Wang [11], Ilic and Notaros [12], and Jorgensen *et al.* [13]. For illustration, we discuss those of [12] in the following. (The bases in [12] are actually curl-conforming functions; for square cell shapes we have converted these to divergence-conforming bases using the simple relation between the two types as discussed in Chapter 4.) Hierarchical vector bases can be organized by their maximum polynomial degree p, and also by their designation as "edge-based" or "cell-based" functions: edge-based functions have a nonzero normal component that must be made continuous with analogous functions in adjacent cells, while cell-based functions are entirely local.

The divergence-conforming basis functions obtained from [12] have the form

$$\hat{u}\, P_i(u) v^j, \quad i = 1, 2, \ldots, N;\, j = 1, 2, \ldots, N - 1 \tag{3.48}$$

$$\hat{v}\, u^i\, P_j(v), \quad i = 1, 2, \ldots, N - 1;\, j = 1, 2, \ldots, N \tag{3.49}$$

where

$$P_i(u) = \begin{cases} 1 - u & i = 0 \\ 1 + u & i = 1 \\ u^i - 1 & i = 2, 4, 6, \ldots \\ u^i - u & i = 3, 5, 7, \ldots \end{cases} \tag{3.50}$$

Table 3.2 presents the first 24 divergence-conforming basis functions, following [12]. The basis functions are not normalized. The first 4 functions are essentially the $p = 0$ functions of (3.3)–(3.6), which are edge-based. The second 4 functions are also edge-based, and serve to elevate the polynomial degree of the representation to a consistently linear behavior in the normal and tangential vector components. The next 4 functions are cell-based, meaning that they provide no normal component on any of the cell edges. The cell-based functions collectively serve to provide a quadratic tangential-vector component along the cell edges. Use of the first 12 basis functions in Table 3.2 provides a representation equivalent to Nedelec's mixed-order $p = 1$ space.

The philosophy underlying the Nedelec spaces suggests that, for a general problem where the leading order derivative in an equation is a divergence operator, the best accuracy will be obtained by consistently using the first 4 degrees of freedom (Nedelec $p = 0$), or the first 12 (Nedelec $p = 1$), or the first 24 (Nedelec $p = 2$). Little improvement is expected by adding the 4 $p = 2$ edge-based basis functions to the set of 12 $p = 1$ functions, since those degrees of freedom will not be balanced

TABLE 3.2: Hierarchical Divergence-Conforming Vector Bases for Quadrilateral Cells

$p = 0$	$\hat{u}(1 - u)$	Four edge-based functions
	$\hat{u}(1 + u)$	
	$\hat{v}(1 - v)$	
	$\hat{v}(1 + v)$	Total degree of freedom = 4
$p = 1$, edge	$\hat{u}(1 - u)v$	Four edge-based functions
	$\hat{u}(1 + u)v$	
	$\hat{v}u(1 - v)$	
	$\hat{v}u(1 + v)$	Total degrees of freedom = 8
$p = 1$, cell	$\hat{u}(u^2 - 1)$	Four cell-based functions
	$\hat{u}(u^2 - 1)v$	
	$\hat{v}(v^2 - 1)$	
	$\hat{v}u(v^2 - 1)$	Total degrees of freedom = 12
$p = 2$, edge	$\hat{u}(1 - u)v^2$	Four edge-based functions
	$\hat{u}(1 + u)v^2$	
	$\hat{v}u^2(1 - v)$	
	$\hat{v}u^2(1 + v)$	Total degrees of freedom = 16
$p = 2$, cell	$\hat{u}(u^2 - 1)v^2$	Eight cell-based functions
	$\hat{u}(u^3 - u)$	
	$\hat{u}(u^3 - u)v$	
	$\hat{u}(u^3 - u)v^2$	
	$\hat{v}u^2(v^2 - 1)$	
	$\hat{v}(v^3 - v)$	
	$\hat{v}u(v^3 - v)$	
	$\hat{v}u^2(v^3 - v)$	Total degrees of freedom = 24

Note. $-1 \leq u, v \leq 1$; basis functions are not normalized; edge-based functions must be made normally continuous with neighboring cells. Adapted from [12].

in the equation. However, while a user of interpolatory basis functions is usually forced to work with an expansion that is either complete to a given polynomial degree or a given Nedelec space, hierarchical functions provide more freedom to adapt the representation to a particular situation. This adaption may be carried out dynamically within a p-refinement algorithm [14].

As previously mentioned, once the coefficients are determined the current density at specific locations is obtained by superimposing the set of hierarchical basis functions (scaled by their coefficients). Thus, the normalization of these basis functions is arbitrary. However, there are times when hierarchical testing functions are used within error estimators to drive adaptive refinement procedures [14]. In that situation, some normalization is usually necessary for the error estimate to be meaningful.

3.10 HIGHER-ORDER HIERARCHICAL FUNCTIONS FOR TRIANGULAR CELLS

Hierarchical basis functions for triangular or tetrahedral cell shapes have been proposed by Andersen and Volakis [15], Webb [16], Preissig and Peterson [17], and others. Most of the published bases are curl-conforming, but these can easily be converted into divergence-conforming functions for triangles using the relation described in Chapter 4. For illustration, Table 3.3 shows the first 24 divergence-conforming basis functions from [17], grouped in order of polynomial degree and separated by classification into "edge-based" and "cell-based" functions. The first three functions are essentially the $p = 0$ functions from (3.12)–(3.14). The next three functions are edge-based and elevate the polynomial degree of the representation to linear. The seventh and eighth entries in the table are cell-based functions that bring the representation up to the Nedelec $p = 1$ space [5]. The edge-based functions must be paired with similar functions in the adjacent cells to provide normal-vector continuity.

TABLE 3.3: Hierarchical Divergence–Conforming Vector Bases for Triangular Cells

Mixed order 0/1	$(u-1)\hat{u} + v\hat{v}$	Three edge-based functions
	$u\hat{u} + (v-1)\hat{v}$	
	$u\hat{u} + v\hat{v}$	Total degrees of freedom = 3
Complete order 1	$u\hat{u} + v\hat{v}$	Three edge-based functions
	$-u\hat{u} + (2u+v-1)\hat{v}$	
	$(1-u-2v)\hat{u} + v\hat{v}$	Total degrees of freedom = 6
Mixed order 1/2	$u[(u-1)\hat{u} + v\hat{v}]$	Two cell-based functions
	$v[u\,\hat{u} + (v-1)\,\hat{v}]$	Total degrees of freedom = 8
Complete order 2	$(u^2 - 2uv)\,\hat{u} - (2uv - v^2)\,\hat{v}$	Three edge-based functions
	$(2uw - u^2)\,\hat{u} - (2uw - [w-u]^2)\,\hat{v}$	
	$(2vw - [w-v]^2)\,\hat{u} - (2vw - v^2)\,\hat{v}$	One cell-based function
	$u(w-v)\,\hat{u} - v(w-u)\,\hat{v}$	Total degrees of freedom = 12
Mixed-order 2/3	$uv[(u-1)\,\hat{u} + v\hat{v}]$	Three cell-based functions
	$uw[(u-1)\,\hat{u} + v\hat{v}]$	
	$vw[u\hat{u} + (v-1)\,\hat{v}]$	Total degrees of freedom = 15

Complete order 3	$2u(u^2 - 5uv + 3v^2)\,\hat{u} - 2v(3u^2 - 5uv + v^2)\,\hat{v}$	Three edge-based functions
	$-2u(u^2 - 5uw + 3w^2)\,\hat{u} - 2(w-u)(u^2 - 7uw + w^2)\,\hat{v}$	Two cell-based functions
	$2(w-v)(v^2 - 7vw + w^2)\,\hat{u} + 2v(v^2 - 5vw + 3w^2)\,\hat{v}$	
	$u[v(v-2w) - u(v-w)]\,\hat{u} - v[u(v-u) + w(v-2u)]\,\hat{v}$	
	$u[u(w-v) - w(w-2v)]\,\hat{u} - v[w(w-4u) - u^2]\,\hat{v}$	Total degrees of freedom = 20
Mixed order 3/4	$uv(u-v)[(u-1)\,\hat{u} + v\,\hat{v}]$	Four cell-based functions
	$uw(u-w)[(u-1)\,\hat{u} + v\,\hat{v}]$	
	$vw(v-w)[u\hat{u} + (v-1)\,\hat{v}]$	
	$uvw[u\hat{u} + (v-1)\hat{v}]$	Total degrees of freedom = 24

Note. Simplex coordinates $u, v, w = 1 - u - v$; $0 \leq u, v, w \leq 1$; basis functions not normalized; edge-based functions must have normal components continuous with neighboring cells. Adapted from [17].

REFERENCES

[1] P. A. Raviart and J. M. Thomas, "Primal hybrid finite element methods for 2nd order elliptic equations," *Math. Comput.*, vol. 31, pp. 391–413, April 1977.

[2] A. W. Glisson, "On the development of numerical techniques for treating arbitrarily shaped surfaces," Ph.D. Dissertation, University of Mississippi, 1978.

[3] A. W. Glisson and D. R. Wilton, "Simple and efficient numerical methods for problems of electromagnetic radiation and scattering from surfaces," *IEEE Trans. Antennas Propagat.*, vol. AP-28, pp. 593–603, Sept. 1980. doi:10.1109/TAP.1980.1142390

[4] S. M. Rao, D. R. Wilton, and A. W. Glisson, "Electromagnetic scattering by surfaces of arbitrary shape," *IEEE Trans. Antennas Propagat.*, vol. AP-30, pp. 409–418, May 1982. doi:10.1109/TAP.1982.1142818

[5] J. C. Nedelec, "Mixed finite elements in R3," *Numer. Math.*, vol. 35, pp. 315–341, 1980. doi:10.1007/BF01396415

[6] J. P. Webb, "Matching a given field using hierarchal vector basis functions," *Electromagnetics*, vol. 24, pp. 113–122, 2004. doi:10.1080/02726340490261590

[7] L. Gurel, K. Sertel, and I. K. Sendur, "On the choice of basis functions to model surface electric current densities in computational electromagnetics," *Radio Sci.*, vol. 34, pp. 1373–1387, Nov./Dec. 1999. doi:10.1029/1999RS900008

[8] A. W. Glisson, S. M. Rao, and D. R. Wilton, "Physically-based approximation of electromagnetic field quantities," in *IEEE Antenn. Propagat. Int. Symp. Digest*, San Antonio, TX, pp. 78–81, July 2002.

[9] R. D. Graglia, D. R. Wilton, and A. F. Peterson, "Higher order interpolatory vector bases for computational electromagnetics," *IEEE Trans. Antenn. Propagat.*, vol. 45, no. 3, pp. 329–342, March 1997. doi:10.1109/8.558649

[10] R. N. Rieben, D. A. White, and G. H. Rodrigue, "Improved conditioning of finite element matrices using new high-order interpolatory bases," *IEEE Trans. Antenn. Propagat.*, vol. 52, pp. 2675–2683, Oct. 2004. doi:10.1109/TAP.2004.834387

[11] J.-S. Wang, "Hierarchic "edge" elements for high-frequency problems," *IEEE Trans. Magnetics*, vol. 33, pp. 1536–1539, March 1997. doi:10.1109/20.582557

[12] M. M. Ilic and B. M. Notaros, "Higher order hierarchical curved hexahedral vector finite elements for electromagnetic modeling," *IEEE Trans. Microwave Theory Tech.*, vol. 51, pp. 1026–1033, March 2003. doi:10.1109/TMTT.2003.808680

[13] E. Jorgensen, J. L. Volakis, P. Meincke, and O. Breinbjerg, "Higher order hierarchical Legendre basis functions for electromagnetic modeling," *IEEE Trans. Antenn. Propagat.*, vol. 52, pp. 2985–2995, Nov. 2004. doi:10.1109/TAP.2004.835279

[14] M. Salazar-Palma, *et al.*, *Iterative and Self-Adaptive Finite Elements in Electromagnetic Modeling.* Boston: Artech House, 1998.

[15] L. S. Andersen and J. L. Volakis, "Hierarchical tangential vector finite elements for tetrahedra," *IEEE Microwave Guided Wave Lett.*, vol. 8, pp. 127–129, March 1998. doi:10.1109/75.661137

[16] J. P. Webb, "Hierarchical vector basis functions of arbitrary order for triangular and tetrahedral finite elements," *IEEE Trans. Antenn. Propagat.*, vol. 47, pp. 1244–1253, August 1999. doi:10.1109/8.791939

[17] R. S. Preissig and A. F. Peterson, "A rationale for p-refinement with the vector Helmholtz equation and two-dimensional vector finite elements," *ACES J.*, vol. 19, pp. 65–75, July 2004.

CHAPTER 4

Curl-Conforming Basis Functions

Curl-conforming basis functions are complementary to the divergence-conforming functions discussed in Chapter 3. Here, curl-conforming functions are considered for square and triangular reference cells. The lowest order functions will be used to discretize the MFIE in Chapter 7. Tables of higher-order functions are included for completeness. The mapping of these functions to curvilinear cells will be the focus of Chapter 5.

4.1 WHAT DOES CURL-CONFORMING MEAN?

The curl of a vector function is

$$
\nabla \times \bar{B} = \hat{x} \left\{ \frac{\partial B_z}{\partial y} - \frac{\partial B_y}{\partial z} \right\} + \hat{y} \left\{ \frac{\partial B_x}{\partial z} - \frac{\partial B_z}{\partial x} \right\} + \hat{z} \left\{ \frac{\partial B_y}{\partial x} - \frac{\partial B_x}{\partial y} \right\} \qquad (4.1)
$$

In a local coordinate system $(\hat{s}, \hat{t}, \hat{n})$, where $\hat{n}$ is normal to the surface, $\hat{n} = \hat{s} \times \hat{t}$, and $\bar{B}$ is a tangential vector function, we may also refer to the surface curl operation

$$
\nabla_s \times \bar{B} = \hat{n} \left\{ \frac{\partial B_t}{\partial s} - \frac{\partial B_s}{\partial t} \right\} \qquad (4.2)
$$

Each component of the curl is proportional to the "twist" of the vector function about a point in the plane perpendicular to that component.

A *curl-conforming* basis function is one that maintains enough continuity to allow it to be differentiated via the curl, while yielding a bounded, well-defined result without the appearance of Dirac delta functions in $\nabla \times \bar{B}$. To achieve this for subsectional representations, the tangential-vector components of $\bar{B}$ must be continuous across cell edges. The normal-vector components are not differentiated across cell edges, and thus cell-to-cell normal-vector continuity is not required. Therefore, *a curl-conforming basis function is one that maintains first-order tangential-vector continuity across cell edges.*

4.2   HISTORY OF THE USE OF CURL-CONFORMING BASIS FUNCTIONS

Curl-conforming basis functions were first proposed for use with finite element solutions by Nedelec [1]. They have been used and extended primarily in connection with solutions of the vector Helmholtz equation (the so-called "curl-curl" form of that equation) in electromagnetic field problems. Early research using lower-order functions were reported by Bossavit and Verite [2], Mur and de Hoop [3], and Barton and Cendes [4], all who used triangular or tetrahedral cells. Crowley employed curl-conforming functions for hexahedral cells [5]. Rao and Wilton appear to have been the first to use curl-conforming basis functions for discretizing integral equations [6].

Interpolatory and hierarchical functions of arbitrary order for square and triangular cell shape have subsequently been proposed. These functions correspond to the Nedelec mixed-order spaces of minimum degree p [1].

4.3   RELATION BETWEEN THE DIVERGENCE-CONFORMING AND CURL-CONFORMING FUNCTIONS

Unlike representations that are mathematically complete to the same polynomial degree in all variables, Nedelec's two-dimensional mixed-order curl-conforming spaces deliberately provide an additional polynomial degree in the variable orthogonal to the vector direction [1]. Because of this behavior, the curl of the basis function is mathematically complete to the same degree as the basis function itself. Equivalently, these spaces discard some of the possible degrees of freedom. When discretizing an equation whose leading-order term is a curl operator, the discarded degrees of freedom are essentially those that do not contribute to a balance of terms in the equation. Consequently, the mixed-order representations are believed to provide a more efficient discretization (fewer unknowns for comparable accuracy).

A similar philosophy was discussed in Chapter 3 in the context of Nedelec's divergence-conforming spaces. In fact, in two dimensions the curl-conforming and divergence-conforming functions are closely related. Given a divergence-conforming function, one can obtain a curl-conforming function of the same Nedelec degree p from the simple operation

$$\hat{n} \times \bar{R}^{\text{div}} = \bar{R}^{\text{curl}} \tag{4.3}$$

where $\hat{n}$ is a unit vector perpendicular to the two-dimensional domain in which the basis functions are defined. It follows that

$$\hat{n} \times \bar{R}^{\text{curl}} = -\bar{R}^{\text{div}} \tag{4.4}$$

Because of this simple relation, curl-conforming functions can immediately be obtained from the divergence-conforming functions already developed in Chapter 3. In the following sections, we employ this relationship as a short-cut to obtain the curl-conforming functions in the reference cells.

4.4   BASIS FUNCTIONS OF ORDER $p = 0$ FOR A SQUARE REFERENCE CELL

Figure 4.1 shows a square reference cell occupying the region $-1 < u < 1, -1 < v < 1$. Four curl-conforming basis functions of the lowest order can be defined within this cell as

$$\bar{R}_1^{\text{curl}} = \frac{u - 1}{2} \hat{v} \tag{4.5}$$

$$\bar{R}_2^{\text{curl}} = \frac{u + 1}{2} \hat{v} \tag{4.6}$$

$$\bar{R}_3^{\text{curl}} = -\frac{v - 1}{2} \hat{u} \tag{4.7}$$

$$\bar{R}_4^{\text{curl}} = -\frac{v + 1}{2} \hat{u} \tag{4.8}$$

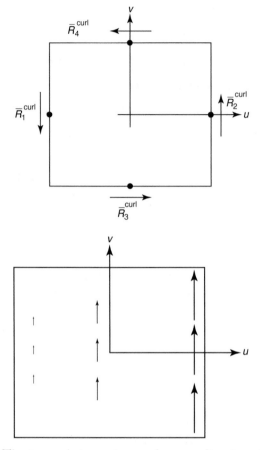

FIGURE 4.1: (a) The interpolation points and vector directions for the $p = 0$ curl-conforming basis functions on a square cell. (b) The function $\bar{R}_2^{\text{curl}}$.

Each of these functions has a nonzero vector component tangential along one edge, and contributes no tangential component along any other edge. For example, the function $\bar{R}_1^{\text{curl}}$ is zero at $u = +1$, and is entirely normal at $v = \pm 1$. Its only nonzero tangential component is along the edge at $u = -1$, and is in the $-\hat{v}$ direction. Each function provides a nonzero tangential component at a different cell edge. It is convenient to think of each as interpolating to the tangential-vector component at the center of that edge.

For these functions to be curl-conforming, they must be matched with basis functions in adjacent cells that maintain the continuity of the nonzero tangential component. Conceptually, similar basis functions are located in cells around those occupied by the preceding functions, for this purpose. The vector direction of the basis functions must be adjusted to maintain a consistent tangential direction across cell boundaries. In addition, the tangential-vector continuity must be maintained by the mapping to curvilinear cells. Curl-conforming functions do not usually maintain the continuity of the normal component of the vector unknown at cell boundaries, and a representation therefore exhibits jump discontinuities in the normal components at cell edges.

The coefficients of the preceding basis functions have an obvious physical interpretation: they are the field or current density component at the cell edge, tangential to that edge. For this interpretation to survive the process of mapping these functions to curvilinear cells, their normalization in the x–y–z space must be controlled. The proper normalization will be considered in connection with the mapping process in Chapter 5.

The reader should compare these curl-conforming basis functions to the divergence-conforming functions given in (3.3)–(3.6), and note that the only difference is the direction of the vector has been rotated by 90° in the u–v plane.

The curl of the preceding functions, carried out in the reference cell coordinates, is given by

$$\nabla \times \bar{R}^{\text{curl}} = (\hat{u} \times \hat{v}) \left\{ \frac{\partial R_v^{\text{curl}}}{\partial u} - \frac{\partial R_u^{\text{curl}}}{\partial v} \right\} \tag{4.9}$$

For the functions in (4.5)–(4.8), the curl within the cell is given by the simple results

$$\frac{\partial R_{1v}^{\text{curl}}}{\partial u} - \frac{\partial R_{1u}^{\text{curl}}}{\partial v} = \frac{1}{2} \tag{4.10}$$

$$\frac{\partial R_{2v}^{\text{curl}}}{\partial u} - \frac{\partial R_{2u}^{\text{curl}}}{\partial v} = \frac{1}{2} \tag{4.11}$$

$$\frac{\partial R_{3v}^{\text{curl}}}{\partial u} - \frac{\partial R_{3u}^{\text{curl}}}{\partial v} = \frac{1}{2} \tag{4.12}$$

$$\frac{\partial R_{4v}^{\text{curl}}}{\partial u} - \frac{\partial R_{4u}^{\text{curl}}}{\partial v} = \frac{1}{2} \tag{4.13}$$

After being mapped to a curvilinear cell in x–y–z space, the curl of the basis functions includes an additional nonconstant scale factor related to the Jacobian of the mapping (Section 5.7). The sign of these results must be adjusted in accordance with any sign change in direction as mentioned above, and any other normalization factor that may be included to achieve a unit component at a desired location within the curvilinear cell.

4.5   BASIS FUNCTIONS OF ORDER $p = 0$ FOR A TRIANGULAR REFERENCE CELL

Figure 4.2 shows a triangular reference cell occupying the region $0 < u < 1, 0 < v < 1, u + v < 1$. Three curl-conforming basis functions of the lowest order can be defined within this cell as

$$\bar{R}_1^{\text{curl}} = v\nabla w - w\nabla v = (u - 1)\hat{v} - v\hat{u} \tag{4.14}$$

$$\bar{R}_2^{\text{curl}} = w\nabla u - u\nabla w = u\hat{v} - (v - 1)\hat{u} \tag{4.15}$$

$$\bar{R}_3^{\text{curl}} = \sqrt{2}(u\nabla v - v\nabla u) = \sqrt{2}(u\hat{v} - v\hat{u}) \tag{4.16}$$

where w is the third simplex coordinate, satisfying $w = 1 - u - v$. Each of these functions interpolates to the vector component tangential to one edge, and contributes no tangential component along any other edge. For example, the function $\bar{R}_1^{\text{curl}}$ contributes a tangential component along the $u = 0$ axis, and is entirely normal or zero along the other two edges. Its tangential component at $u = 0$ is in the $-\hat{v}$ direction. Function $\bar{R}_2^{\text{curl}}$ provides a tangential component along the $v = 0$ axis, in the $+\hat{u}$ direction, while $\bar{R}_3^{\text{curl}}$ contributes a tangential component only along the edge at $u + v = 1$. These functions are the same as the three divergence-conforming basis functions of (3.12)–(3.14), with their vector directions rotated by $90°$ in the u–v plane.

For these functions to be curl-conforming, they must be matched with basis functions in adjacent cells that maintain the continuity of the tangential component. In addition, the vector direction of the basis function in cells sharing an edge must be adjusted to maintain a continuous tangential component across that edge.

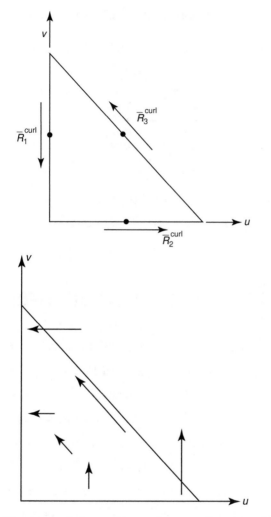

FIGURE 4.2: (a) The interpolation points and vector directions for the $p = 0$ curl-conforming basis functions on the unit triangle. (b) The function $\bar{R}_3^{\text{curl}}$.

The continuity of the normal-vector components is not maintained across cell boundaries.

As in the square cell case, the coefficients of the preceding basis functions have the physical interpretation that they are the tangential component of the field or current density at each cell edge. The normalization necessary to achieve a unit component after mapping is considered in Chapter 5.

The curl of these functions in the local coordinates of the reference cell is obtained as

$$\frac{\partial R_{1v}^{\text{curl}}}{\partial u} - \frac{\partial R_{1u}^{\text{curl}}}{\partial v} = 2 \tag{4.17}$$

$$\frac{\partial R_{2v}^{\text{curl}}}{\partial u} - \frac{\partial R_{2u}^{\text{curl}}}{\partial v} = 2 \tag{4.18}$$

$$\frac{\partial R_{3v}^{\text{curl}}}{\partial u} - \frac{\partial R_{3u}^{\text{curl}}}{\partial v} = 2\sqrt{2} \tag{4.19}$$

As explained in Chapter 5, the curl of the basis functions in the x–y–z space is different due to the presence of a nonconstant scale factor, possible sign change, and normalization constant.

4.6   HIGHER-ORDER INTERPOLATORY FUNCTIONS FOR SQUARE CELLS

Curl-conforming functions may be constructed for square cells using the same arrangement of interpolation points depicted in Fig. 3.5 for the 12 Nedelec $p = 1$ functions, or Fig. 3.6 for the 24 $p = 2$ functions, based on [7]. It is convenient to employ shifted Silvester polynomials to generate the higher-order functions, as explained in Section 3.6 for divergence conforming bases. Table 4.1 presents the functions for degrees $p = 0$, $p = 1$, and $p = 2$. The functions labeled "edge-based" must be paired with analogous functions in adjacent cells in order to maintain tangential-vector continuity. The "cell-based" functions are entirely local and contribute no tangential component on any of the cell edges. As described in [7], the procedure can be continued to obtain bases of any degree.

4.7   HIGHER-ORDER INTERPOLATORY FUNCTIONS FOR TRIANGULAR CELLS

Figure 3.7 presented triangular-cell interpolation points that can be used to construct Nedelec $p = 1$ and $p = 2$ functions, also following the approach of [7]. The construction of curl-conforming basis functions is similar to that carried out

TABLE 4.1: Interpolatory Curl-Conforming Vector Bases for Square Cells

$p = 0$	$\dfrac{u-1}{2}\hat{v}$	Four edge-based functions
	$\dfrac{u+1}{2}\hat{v}$	
	$-\dfrac{v-1}{2}\hat{u}$	Total degrees of freedom $= 4$
	$-\dfrac{v+1}{2}\hat{u}$	
$p = 1$, edge	$-\hat{v}\dfrac{u(u-1)(3v+1)}{4}$	Eight edge-based functions
	$\hat{v}\dfrac{u(u+1)(3v+1)}{4}$	
	$-\hat{v}\dfrac{u(u-1)(1-3v)}{4}$	
	$\hat{v}\dfrac{u(u+1)(1-3v)}{4}$	
	$\hat{u}\dfrac{(1-3u)v(v-1)}{4}$	
	$-\hat{u}\dfrac{(1-3u)v(v+1)}{4}$	
	$\hat{u}\dfrac{(3u+1)v(v-1)}{4}$	
	$-\hat{u}\dfrac{(3u+1)v(v+1)}{4}$	
$p = 1$, cell	$\hat{v}\dfrac{(1-u)(1+u)(3v+1)}{2}$	Four cell-based functions
	$\hat{v}\dfrac{(1-u)(1+u)(1-3v)}{2}$	
	$-\hat{u}\dfrac{(1-3u)(1-v)(1+v)}{2}$	
	$-\hat{u}\dfrac{(3u+1)(1-v)(1+v)}{2}$	Total degrees of freedom $= 12$

(Continued)

TABLE 4.1: (*Continued*)

$p = 2$, edge	$\hat{u}\,u(1-2u)(1+3v)(1-3v)(1-v)/16$	Twelve edge-based functions
	$-\hat{u}\,(1-2u)(1+2u)(1+3v)(1-3v)(1-v)/16$	
	$-\hat{u}\,u(1+2u)(1+3v)(1-3v)(1-v)/16$	
	$\hat{v}\,(1+u)(1+3u)(1-3u)v(1-2v)/16$	
	$-\hat{v}\,(1+u)(1+3u)(1-3u)(1-2v)(1+2v)/16$	
	$-\hat{v}\,(1+u)(1+3u)(1-3u)v(1+2v)/16$	
	$-\hat{u}\,u(1-2u)(1+v)(1+3v)(1-3v)/16$	
	$\hat{u}\,(1-2u)(1+2u)(1+v)(1+3v)(1-3v)/16$	
	$\hat{u}u(1+2u)(1+v)(1+3v)(1-3v)/16$	
	$\hat{v}\,(1+3u)(1-3u)(1-u)v(1+2v)/16$	
	$\hat{v}\,(1+3u)(1-3u)(1-u)(1-2v)(1+2v)/16$	
	$-\hat{v}\,(1+3u)(1-3u)(1-u)v(1-2v)/16$	
$p = 2$, cell	$\hat{u}u(1-2u)(1-3v)(1-v^2)$	Twelve cell-based functions
	$\hat{u}\,(1-2u)(1+2u)(1-3v)(1-v^2)$	
	$\hat{u}\,u(1+2u)(1-3v)(1-v^2)$	
	$\hat{u}\,u(1-2u)(1+3v)(1-v^2)$	
	$\hat{u}\,u(1-2u)(1+2u)(1+3v)(1-v^2)$	
	$\hat{u}\,u(1+2u)(1+3v)(1-v^2)$	
	$\hat{v}\,(1-3u)(1-u^2)v(1-2v)$	
	$\hat{v}\,(1-3u)(1-u^2)(1-2v)(1+2v)$	
	$\hat{v}\,(1-3u)(1-u^2)v(1+2v)$	
	$\hat{v}\,(1+3u)(1-u^2)v(1-2v)$	
	$\hat{v}\,(1+3u)(1-u^2)(1-2v)(1+2v)$	
	$\hat{v}\,(1+3u)(1-u^2)v(1+2v)$	Total degrees of freedom $= 24$

Note. $-1 \le u, v \le 1$; basis functions normalized to unity; edge-based functions must be made normally continuous with neighboring cells. After [7].

in Chapter 3 for divergence-conforming functions, by forming products of the $p = 0$ functions with appropriate shifted Silvester polynomials. For example, the functions that interpolate to nodes labeled 103, 202, and 301 in Fig. 3.7(b) are constructed from

$$\bar{R}_{103} = \bar{R}_2^{\text{curl}} S_3^{(4)}(w) \tag{4.20}$$
$$\bar{R}_{202} = \bar{R}_2^{\text{curl}} S_2^{(4)}(u) S_2^{(4)}(w) \tag{4.21}$$
$$\bar{R}_{301} = \bar{R}_2^{\text{curl}} S_3^{(4)}(u) \tag{4.22}$$

where w is the third simplex coordinate, and the shifted Silvester polynomial is defined in (3.19). Table 4.2 presents the $p = 0$, $p = 1$, and $p = 2$ functions. Tangential-vector continuity is maintained by pairing the functions labeled "edge-based" with analogous functions in adjacent cells.

4.8   HIGHER-ORDER HIERARCHICAL FUNCTIONS FOR SQUARE CELLS

Hierarchical functions for quadrilateral or hexahedral cells were summarized in Chapter 3, in the context of divergence-conforming bases. Most of the functions reported in the literature are actually curl-conforming, intended for the finite element solution of the curl-curl form of the vector Helmholtz equation. For illustration, we consider the curl-conforming bases of Ilic and Notaros [8], which have the form

$$\hat{u} u^i P_j(v), \quad i = 1, 2, \ldots, N - 1; j = 1, 2, \ldots, N \tag{4.23}$$
$$\hat{v} P_i(u) v^j, \quad i = 1, 2, \ldots, N; j = 1, 2, \ldots, N - 1 \tag{4.24}$$

where P_i is defined in Eq. (3.50). Table 4.3 presents the first 24 basis functions, organized by Nedelec degree and by the edge-based or cell-based nature of the functions. The cell-based functions are entirely local to a cell and have no tangential components along the cell boundaries. The edge-based functions must be paired with similar functions in adjacent cells to ensure tangential-vector continuity.

TABLE 4.2: Interpolatory Curl-Conforming Vector Bases for Triangular Cells

Mixed order	$(u-1)\hat{v} - v\hat{u}$	Three edge-based
0/1: $(p=0)$	$u\hat{v} - (v-1)\hat{u}$	functions
	$\sqrt{2}(u\hat{v} - v\hat{u})$	Total degrees of
		freedom $= 3$
Mixed order	$\bar{R}_{012} = \{(u-1)\hat{v} - v\hat{u}\}(3w-1)$	Six edge-based
1/2: $(p=1)$	$\bar{R}_{021} = \{(u-1)\hat{v} - v\hat{u}\}(3v-1)$	functions
	$\bar{R}_{102} = \{u\hat{v} - (v-1)\hat{u}\}(3w-1)$	
	$\bar{R}_{201} = \{u\hat{v} - (v-1)\hat{u}\}(3u-1)$	
	$\bar{R}_{120} = \{\sqrt{2}(u\hat{v} - v\hat{u}\}(3v-1)$	
	$\bar{R}_{210} = \{\sqrt{2}(u\hat{v} - v\hat{u})\}(3u-1)$	
	$\bar{R}_{111a} = \{(u-1)\hat{v} - v\hat{u}\}u$	Two cell-based
		functions
	$\bar{R}_{111b} = \{u\hat{v} - (v-1)\hat{u}\}v$	Total degrees of
		freedom $= 8$
Mixed-order	$\bar{R}_{013} = \{(u-1)\hat{v} - v\hat{u}\}(4w-1)(4w-2)/2$	Nine edge-based
2/3: $(p=2)$	$\bar{R}_{022} = \{(u-1)\hat{v} - v\hat{u}\}(4v-1)(4w-1)$	functions
	$\bar{R}_{031} = \{(u-1)\hat{v} - v\hat{u}\}(4v-1)(4v-2)/2$	
	$\bar{R}_{103} = \{u\hat{v} - (v-1)\hat{u}\}(4w-1)(4w-2)/2$	
	$\bar{R}_{202} = \{u\hat{v} - (v-1)\hat{u}\}(4u-1)(4w-1)$	
	$\bar{R}_{301} = \{u\hat{v} - (v-1)\hat{u}\}(4u-1)(4u-2)/2$	
	$\bar{R}_{130} = \{\sqrt{2}(u\hat{v} - v\hat{u})\}(4v-1)(4v-2)/2$	
	$\bar{R}_{220} = \{\sqrt{2}(u\hat{v} - v\hat{u})\}(4u-1)(4v-1)$	
	$\bar{R}_{310} = \{\sqrt{2}(u\hat{v} - v\hat{u})\}(4u-1)(4u-2)/2$	
	$\bar{R}_{112a} = \{(u-1)\hat{v} - v\hat{u}\}u(4w-1)$	Six cell-based
		functions

TABLE 4.2: (*Continued*)

$$\bar{R}_{112b} = \{u\hat{v} - (v-1)\hat{u}\}v(4w-1)$$
$$\bar{R}_{121a} = \{(u-1)\hat{v} - v\hat{u}\}u(4v-1)$$
$$\bar{R}_{121b} = \{u\hat{v} - (v-1)\hat{u}\}v(4v-1)$$
$$\bar{R}_{211a} = \{(u-1)\hat{v} - v\hat{u}\}u(4u-1)$$
$$\bar{R}_{211b} = \{u\hat{v} - (v-1)\hat{u}\}v(4u-1)$$

Total degrees of
freedom $= 15$

Note. Simplex coordinates u, v, $w = 1 - u - v$; $0 \leq u$, v, $w \leq 1$; edge basis functions normalized to unity; edge-based functions must have tangential components continuous with neighboring cells. After [7].

In contrast to the organization of Tables 4.1 and 4.2, the functions in Table 4.3 are intended to build upon the lower-order bases. Thus, the full $p = 2$ representation must include the $p = 0$ and $p = 1$ functions in the table.

4.9 HIGHER-ORDER HIERARCHICAL FUNCTIONS FOR TRIANGULAR CELLS

The literature contains a variety of hierarchical bases for triangular or tetrahedral cells. Table 4.4 summarizes the first 24 curl-conforming basis functions from [9], organized by Nedelec degree and by their edge-based or cell-based nature. The edge-based functions must be paired with similar functions in adjacent cells to provide tangential-vector continuity. A subset of the triangular basis functions of any mixed-order Nedelec degree p form a polynomial-complete set of degree p, and this subset is also noted in the table. The full representation of any degree includes all the lower-degree functions listed in Table 4.4.

TABLE 4.3: Hierarchical Curl-Conforming Vector Bases for Square Cells

$p = 0$	$\hat{u}(1 - v)$	Four edge-based functions
	$\hat{u}(1 + v)$	
	$\hat{v}(1 - u)$	
	$\hat{v}(1 + u)$	Total degrees of freedom $= 4$
$p = 1$, edge	$\hat{u}u(1 - v)$	Four edge-based functions
	$\hat{u}\,u(1 + v)$	
	$\hat{v}(1 - u)v$	
	$\hat{v}(1 + u)v$	Total degrees of freedom $= 8$
$p = 1$, cell	$\hat{u}(v^2 - 1)$	Four cell-based functions
	$\hat{u}\,u(v^2 - 1)$	
	$\hat{v}(u^2 - 1)$	
	$\hat{v}(u^2 - 1)v$	Total degrees of freedom $= 12$
$p = 2$, edge	$\hat{u}u^2(1 - v)$	Four edge-based functions
	$\hat{u}\,u^2(1 + v)$	
	$\hat{v}(1 - u)v^2$	
	$\hat{v}(1 + u)v^2$	Total degrees of freedom $= 16$
$p = 2$, cell	$\hat{u}u^2(v^2 - 1)$	Eight cell-based functions
	$\hat{u}\,(v^3 - v)$	
	$\hat{u}\,u(v^3 - v)$	
	$\hat{u}\,u^2(v^3 - v)$	
	$\hat{v}(u^2 - 1)v^2$	
	$\hat{v}(u^3 - u)$	
	$\hat{v}(u^3 - u)v$	
	$\hat{v}(u^3 - u)v^2$	Total degrees of freedom $= 24$

Note. $-1 \leq u, v \leq 1$; basis functions are not normalized; edge-based functions must be made tangentially continuous with neighboring cells. After [8].

TABLE 4.4: Hierarchical Curl-Conforming Vector Bases for Triangular Cells		
Mixed order 0/1: ($p = 0$)	$v\nabla w - w\nabla v$	Three edge-based functions
	$w\nabla u - u\nabla w$	
	$u\nabla v - v\nabla u$	Total degrees of freedom = 3
Complete degree 1	$\nabla(uv)$	Three edge-based functions
	$\nabla(uw)$	
	$\nabla(vw)$	Total degrees of freedom = 6
Mixed order 1/2: ($p = 1$)	$u(v\nabla w - w\nabla v)$	Two cell-based functions
	$v(w\nabla u - u\nabla w)$	Total degrees of freedom = 8
Complete degree 2	$\nabla\{uv(u - v)\}$	Three edge-based functions
	$\nabla\{uw(u - w)\}$	One cell-based function
	$\nabla\{vw(v - w)\}$	
	$\nabla\{uvw\}$	Total degrees of freedom = 12
Mixed-order 2/3: ($p = 2$)	$uv(v\nabla w - w\nabla v)$	Three cell-based functions
	$uw(v\nabla w - w\nabla v)$	
	$vw(w\nabla u - u\nabla w)$	Total degrees of freedom = 15
		(*Continued*)

TABLE 4.4: (*Continued*)		
Complete degree 3	$\nabla\{uv(2u - v)(u - 2v)\}$	Three edge-based functions, two cell-based functions
	$\nabla\{uw(2u - w)(u - 2w)\}$	
	$\nabla\{vw(2v - w)(v - 2w)\}$	
	$\nabla\{uvw \ (u - v)\}$	
	$\nabla\{uvw(u - w)\}$	Total degrees of freedom = 20
Mixed-order 3/4: ($p = 3$)	$uv(u - v)(v\nabla w - w\nabla v)$	Four cell-based functions
	$uw(u - w)(v\nabla w - w\nabla v)$	
	$vw(v - w)(w\nabla u - u\nabla w)$	
	$uvw(w\nabla u - u\nabla w)$	Total degrees of freedom = 24

Note. Simplex coordinates $u, v, w = 1 - u - v; 0 \leq u, v, w \leq 1$; basis functions not normalized; edge-based functions must have tangential components continuous with neighboring cells. After [9].

REFERENCES

[1] J. C. Nedelec, "Mixed finite elements in R3," *Numer. Math.*, vol. 35, pp. 315–341, 1980. doi:10.1007/BF01396415

[2] A. Bossavit and J. C. Verite, "A mixed FEM-BIEM method to solve 3D eddy current problems," *IEEE Trans. Magnetics*, vol. MAG-18, pp. 431–435, March 1982. doi:10.1109/TMAG.1982.1061847

[3] G. Mur and A. T. de Hoop, "A finite element method for computing three-dimensional electromagnetic fields in inhomogeneous media," *IEEE Trans. Magnetics*, vol. MAG-21, pp. 2188–2191, Nov. 1985. doi:10.1109/TMAG.1985.1064256

[4]   M. L. Barton and Z. J. Cendes, "New vector finite elements for three-dimensional magnetic field computation," *J. Appl. Phys.*, vol. 61, pp. 3919–3921, April 1987. doi:10.1063/1.338584

[5]   C. W. Crowley, "Mixed-order covariant projection finite elements for vector fields," Ph.D. Dissertation, McGill University, Montreal, Quebec, Feb. 1988.

[6]   S. M. Rao and D. R. Wilton, "E-field, H-field, and combined-field solution for arbitrarily shaped three-dimensional dielectric bodies," *Electromagnetics*, vol. 10, pp. 407–421, 1990.

[7]   R. D. Graglia, D. R. Wilton, and A. F. Peterson, "Higher order interpolatory vector bases for computational electromagnetics," *IEEE Trans. Antenn. Propagat.*, vol. 45, no. 3, pp. 329–342, March 1997. doi:10.1109/8.558649

[8]   M. M. Ilic and B. M. Notaros, "Higher order hierarchical curved hexahedral vector finite elements for electromagnetic modeling," *IEEE Trans. Microwave Theory Tech.*, vol. 51, pp. 1026–1033, March 2003. doi:10.1109/TMTT.2003.808680

[9]   R. S. Preissig and A. F. Peterson, "A rationale for p-refinement with the vector Helmholtz equation and two-dimensional vector finite elements," *ACES J.*, vol. 19, pp. 65–75, July 2004.

CHAPTER 5

Transforming Vector Basis Functions to Curved Cells

This chapter explains the steps required to map a vector basis function from a 2D reference cell to a curvilinear surface in x–y–z space. To begin in a logical manner, we first consider a mapping from 3-space to 3-space. The process of mapping vector functions is somewhat more complicated than the scalar mapping used in Chapter 2 to define the cell shape and location. For this reason, it is necessary to introduce base vectors, reciprocal base vectors, and a number of other vector relations. After the basis function transformation is discussed, we consider the conversion of vector derivatives on the curved cells to equivalent derivatives in the reference coordinates.

Vector parametric mappings were described by Stratton in his 1941 textbook [1], and will be reviewed below in a slightly different notation. Several curved-cell implementations that used the same type of vector mapping described here may be studied for reference [2–5]. These approaches will be used in Chapters 6 and 7 for the numerical solution of electromagnetic integral equations.

5.1 BASE VECTORS AND RECIPROCAL BASE VECTORS

Suppose we have the general three-dimensional situation involving a mapping from a 3D reference space described by coordinates (u, v, w)[1] to the x–y–z space. For

---

[1] In previous chapters, w was used as the third simplex coordinate in 2D space. Here, it will be used for the third dimension in the reference space.

our purposes, it is convenient to assume that the mapping functions $x(u, v, w)$, $y(u, v, w)$, and $z(u, v, w)$ are defined on a cell-by-cell basis using Lagrangian interpolation polynomials (as described in Chapter 2, but generalized to a 3D reference cell). A position vector from the origin $(0, 0, 0)$ to a point (x, y, z) in the curved cell is expressed

$$\bar{r}(u, v, w) = x(u, v, w)\hat{x} + y(u, v, w)\hat{y} + z(u, v, w)\hat{z} \tag{5.1}$$

From a study of the differential displacement

$$d\bar{r} = \frac{\partial \bar{r}}{\partial u}du + \frac{\partial \bar{r}}{\partial v}dv + \frac{\partial \bar{r}}{\partial w}dw \tag{5.2}$$

it is apparent that three displacement vectors can be defined as

$$\bar{s} = \frac{\partial \bar{r}}{\partial u} = \frac{\partial x}{\partial u}\hat{x} + \frac{\partial y}{\partial u}\hat{y} + \frac{\partial z}{\partial u}\hat{z} \tag{5.3}$$

$$\bar{t} = \frac{\partial \bar{r}}{\partial v} = \frac{\partial x}{\partial v}\hat{x} + \frac{\partial y}{\partial v}\hat{y} + \frac{\partial z}{\partial v}\hat{z} \tag{5.4}$$

$$\bar{n} = \frac{\partial \bar{r}}{\partial w} = \frac{\partial x}{\partial w}\hat{x} + \frac{\partial y}{\partial w}\hat{y} + \frac{\partial z}{\partial w}\hat{z} \tag{5.5}$$

If parameters v and w are held constant, while u is varied, the mapping creates a curve. The vector $\bar{s}$ is tangential to that curve. Similarly, $\bar{t}$ is tangential to a curve defined by constant values of parameters u and w, and $\bar{n}$ is tangential to a curve defined by constant values of u and v. These three vectors are known as *base vectors*. If all three parameters are varied between constant limits to create a curvilinear cell in x–y–z space, two of the three base vectors are tangential to each face of that curvilinear cell. The base vectors are not necessarily mutually perpendicular at a point, nor are they unit vectors in general.

Alternatively, we can define three independent vectors in terms of the gradients

$$\bar{s}' = \nabla u = \frac{\partial u}{\partial x}\hat{x} + \frac{\partial u}{\partial y}\hat{y} + \frac{\partial u}{\partial z}\hat{z} \tag{5.6}$$

$$\bar{t}' = \nabla v = \frac{\partial v}{\partial x}\hat{x} + \frac{\partial v}{\partial y}\hat{y} + \frac{\partial v}{\partial z}\hat{z} \tag{5.7}$$

$$\bar{n}' = \nabla w = \frac{\partial w}{\partial x}\hat{x} + \frac{\partial w}{\partial y}\hat{y} + \frac{\partial w}{\partial z}\hat{z} \tag{5.8}$$

It should be apparent from the gradient operation that the vector $\bar{s}'$ is normal to a surface over which u is constant, while $\bar{t}'$ is normal to a surface on which v is constant. Similarly, the vector $\bar{n}'$ is normal to a surface of constant w. These vectors are known as *reciprocal base vectors*. The reciprocal base vectors are also not necessarily mutually perpendicular or of unit length. If parameters u, v, and w are varied between constant limits to create a curvilinear cell in $x–y–z$ space, one reciprocal base vector is normal at every point on each face of the resulting curvilinear cell.

5.2   JACOBIAN RELATIONS

In the mapping between the reference cell and the curvilinear cell in $x–y–z$ space, derivatives transform according to the relation

$$\begin{bmatrix} \dfrac{\partial}{\partial u} \\[2ex] \dfrac{\partial}{\partial v} \\[2ex] \dfrac{\partial}{\partial w} \end{bmatrix} = \begin{bmatrix} \dfrac{\partial x}{\partial u} & \dfrac{\partial y}{\partial u} & \dfrac{\partial z}{\partial u} \\[2ex] \dfrac{\partial x}{\partial v} & \dfrac{\partial y}{\partial v} & \dfrac{\partial z}{\partial v} \\[2ex] \dfrac{\partial x}{\partial w} & \dfrac{\partial y}{\partial w} & \dfrac{\partial z}{\partial w} \end{bmatrix} \begin{bmatrix} \dfrac{\partial}{\partial x} \\[2ex] \dfrac{\partial}{\partial y} \\[2ex] \dfrac{\partial}{\partial z} \end{bmatrix} \tag{5.9}$$

The three-by-three matrix in (5.9) is known as the Jacobian matrix

$$\mathbf{J} = \begin{bmatrix} \dfrac{\partial x}{\partial u} & \dfrac{\partial y}{\partial u} & \dfrac{\partial z}{\partial u} \\[2ex] \dfrac{\partial x}{\partial v} & \dfrac{\partial y}{\partial v} & \dfrac{\partial z}{\partial v} \\[2ex] \dfrac{\partial x}{\partial w} & \dfrac{\partial y}{\partial w} & \dfrac{\partial z}{\partial w} \end{bmatrix} \tag{5.10}$$

The differential volumes of the two spaces are related by the determinant of the Jacobian matrix

$$dx\,dy\,dz = (\det \mathbf{J})du\,dv\,dw \tag{5.11}$$

It is also useful to consider the inverse relation

$$
\begin{bmatrix} \dfrac{\partial}{\partial x} \\[2mm] \dfrac{\partial}{\partial y} \\[2mm] \dfrac{\partial}{\partial z} \end{bmatrix} = \mathbf{J}^{-1} \begin{bmatrix} \dfrac{\partial}{\partial u} \\[2mm] \dfrac{\partial}{\partial v} \\[2mm] \dfrac{\partial}{\partial w} \end{bmatrix} \tag{5.12}
$$

where the inverse of the Jacobian matrix is given directly in the form

$$
\mathbf{J}^{-1} = \begin{bmatrix} \dfrac{\partial u}{\partial x} & \dfrac{\partial v}{\partial x} & \dfrac{\partial w}{\partial x} \\[3mm] \dfrac{\partial u}{\partial y} & \dfrac{\partial v}{\partial y} & \dfrac{\partial w}{\partial y} \\[3mm] \dfrac{\partial u}{\partial z} & \dfrac{\partial v}{\partial z} & \dfrac{\partial w}{\partial z} \end{bmatrix} \tag{5.13}
$$

Observe that the rows of $\mathbf{J}$ are the components of the base vectors, while the columns of $\mathbf{J}^{-1}$ are those of the reciprocal base vectors. Of course, a matrix and its inverse are also related by

$$
\mathbf{J}\mathbf{J}^{-1} = \mathbf{I} \tag{5.14}
$$

where $\mathbf{I}$ is the identity matrix. Embodied in (5.14) are the nine equations

$$
\frac{\partial x}{\partial u}\frac{\partial u}{\partial x} + \frac{\partial y}{\partial u}\frac{\partial u}{\partial y} + \frac{\partial z}{\partial u}\frac{\partial u}{\partial z} = 1 \tag{5.15}
$$

$$
\frac{\partial x}{\partial v}\frac{\partial v}{\partial x} + \frac{\partial y}{\partial v}\frac{\partial v}{\partial y} + \frac{\partial z}{\partial v}\frac{\partial v}{\partial z} = 1 \tag{5.16}
$$

$$
\frac{\partial x}{\partial w}\frac{\partial w}{\partial x} + \frac{\partial y}{\partial w}\frac{\partial w}{\partial y} + \frac{\partial z}{\partial w}\frac{\partial w}{\partial z} = 1 \tag{5.17}
$$

$$
\frac{\partial x}{\partial u}\frac{\partial v}{\partial x} + \frac{\partial y}{\partial u}\frac{\partial v}{\partial y} + \frac{\partial z}{\partial u}\frac{\partial v}{\partial z} = 0 \tag{5.18}
$$

$$
\frac{\partial x}{\partial u}\frac{\partial w}{\partial x} + \frac{\partial y}{\partial u}\frac{\partial w}{\partial y} + \frac{\partial z}{\partial u}\frac{\partial w}{\partial z} = 0 \tag{5.19}
$$

$$
\frac{\partial x}{\partial v}\frac{\partial u}{\partial x} + \frac{\partial y}{\partial v}\frac{\partial u}{\partial y} + \frac{\partial z}{\partial v}\frac{\partial u}{\partial z} = 0 \tag{5.20}
$$

$$
\frac{\partial x}{\partial v}\frac{\partial w}{\partial x} + \frac{\partial y}{\partial v}\frac{\partial w}{\partial y} + \frac{\partial z}{\partial v}\frac{\partial w}{\partial z} = 0 \tag{5.21}
$$

$$\frac{\partial x}{\partial w}\frac{\partial u}{\partial x} + \frac{\partial y}{\partial w}\frac{\partial u}{\partial y} + \frac{\partial z}{\partial w}\frac{\partial u}{\partial z} = 0 \tag{5.22}$$

$$\frac{\partial x}{\partial w}\frac{\partial v}{\partial x} + \frac{\partial y}{\partial w}\frac{\partial v}{\partial y} + \frac{\partial z}{\partial w}\frac{\partial v}{\partial z} = 0 \tag{5.23}$$

From the definition of the base vectors and reciprocal base vectors, we observe that Eqs. (5.15)–(5.17) are equivalent to

$$\bar{s} \bullet \bar{s}' = 1 \tag{5.24}$$

$$\bar{t} \bullet \bar{t}' = 1 \tag{5.25}$$

and

$$\bar{n} \bullet \bar{n}' = 1 \tag{5.26}$$

Equations (5.18)–(5.23) are equivalent to

$$\bar{s} \bullet \bar{t}' = 0 \tag{5.27}$$

$$\bar{s} \bullet \bar{n}' = 0 \tag{5.28}$$

$$\bar{t} \bullet \bar{s}' = 0 \tag{5.29}$$

$$\bar{t} \bullet \bar{n}' = 0 \tag{5.30}$$

$$\bar{n} \bullet \bar{s}' = 0 \tag{5.31}$$

$$\bar{n} \bullet \bar{t}' = 0 \tag{5.32}$$

Thus, the base vectors and reciprocal base vectors satisfy these orthogonality relations.

To further explore the interrelation of the base and reciprocal base vectors, consider a constant-w surface, with base vectors $\bar{s}$ and $\bar{t}$ tangential to that surface. Their cross product can be expanded to obtain

$$\begin{aligned}
\bar{s} \times \bar{t} &= \frac{\partial \bar{r}}{\partial u} \times \frac{\partial \bar{r}}{\partial v} \\
&= \left\{ \frac{\partial x}{\partial u}\hat{x} + \frac{\partial y}{\partial u}\hat{y} + \frac{\partial z}{\partial u}\hat{z} \right\} \times \left\{ \frac{\partial x}{\partial v}\hat{x} + \frac{\partial y}{\partial v}\hat{y} + \frac{\partial z}{\partial v}\hat{z} \right\} \\
&= \hat{x}\left\{ \frac{\partial y}{\partial u}\frac{\partial z}{\partial v} - \frac{\partial y}{\partial v}\frac{\partial z}{\partial u} \right\} + \hat{y}\left\{ \frac{\partial z}{\partial u}\frac{\partial x}{\partial v} - \frac{\partial z}{\partial v}\frac{\partial x}{\partial u} \right\} \\
&\quad + \hat{z}\left\{ \frac{\partial x}{\partial u}\frac{\partial y}{\partial v} - \frac{\partial x}{\partial v}\frac{\partial y}{\partial u} \right\}
\end{aligned} \tag{5.33}$$

This vector must point in a direction normal to the surface of constant w. However, the reciprocal base vector $\bar{n}'$ also points normal to a constant-w surface. That suggests that

$$K\bar{n}' = \bar{s} \times \bar{t} \tag{5.34}$$

for some scalar function $K(u, v, w)$. From a consideration of the determinant of the Jacobian matrix, it is apparent that

$$
\begin{aligned}
\det \mathbf{J} &= \frac{\partial x}{\partial w} \left\{ \frac{\partial y}{\partial u} \frac{\partial z}{\partial v} - \frac{\partial y}{\partial v} \frac{\partial z}{\partial u} \right\} + \frac{\partial y}{\partial w} \left\{ \frac{\partial z}{\partial u} \frac{\partial x}{\partial v} - \frac{\partial z}{\partial v} \frac{\partial x}{\partial u} \right\} \\
&\quad + \frac{\partial z}{\partial w} \left\{ \frac{\partial x}{\partial u} \frac{\partial y}{\partial v} - \frac{\partial x}{\partial v} \frac{\partial y}{\partial u} \right\} \\
&= \frac{\partial \bar{r}}{\partial w} \bullet \frac{\partial \bar{r}}{\partial u} \times \frac{\partial \bar{r}}{\partial v} \\
&= \bar{n} \bullet \bar{s} \times \bar{t}
\end{aligned}
\tag{5.35}
$$

As an aside, we note that the Jacobian determinant can be written in several equivalent forms:

$$\det \mathbf{J} = \bar{n} \bullet \bar{s} \times \bar{t} = \bar{s} \bullet \bar{t} \times \bar{n} = \bar{t} \bullet \bar{n} \times \bar{s} \tag{5.36}$$

It follows from (5.34) and (5.35) that

$$\bar{n} \bullet (K\bar{n}') = \bar{n} \bullet (\bar{s} \times \bar{t}) = \det \mathbf{J} \tag{5.37}$$

However, since $\bar{n} \bullet \bar{n}' = 1$, we obtain the result that the function K is given by

$$K = \det \mathbf{J} \tag{5.38}$$

and we conclude from (5.34) that, at all points within the curvilinear cell,

$$\bar{n}' = \frac{1}{\det \mathbf{J}} \bar{s} \times \bar{t} \tag{5.39}$$

Similarly, the other reciprocal base vectors can be written as

$$\bar{s}' = \frac{1}{\det \mathbf{J}} \bar{t} \times \bar{n} \tag{5.40}$$

$$\bar{t}' = \frac{1}{\det \mathbf{J}} \bar{n} \times \bar{s} \tag{5.41}$$

It is easy to show that

$$\det\left(\mathbf{J}^{-1}\right) = \frac{1}{\det \mathbf{J}} = \bar{s}' \bullet \bar{t}' \times \bar{n}' = \bar{t}' \bullet \bar{n}' \times \bar{s}' = \bar{n}' \bullet \bar{s}' \times \bar{t}' \tag{5.42}$$

and it follows by analogous reasoning that the base vectors can be expressed as

$$\bar{s} = \det \mathbf{J}\left(\bar{t}' \times \bar{n}'\right) \tag{5.43}$$

$$\bar{t} = \det \mathbf{J}\left(\bar{n}' \times \bar{s}'\right) \tag{5.44}$$

$$\bar{n} = \det \mathbf{J}\left(\bar{s}' \times \bar{t}'\right) \tag{5.45}$$

5.3   REPRESENTATION OF VECTOR FIELDS

An understanding of the base and reciprocal base vectors is critical for the representation of vector quantities, such as the surface current density or the electric field within some region. If represented in terms of projections onto the base vectors, known as *covariant* components, one obtains the expression

$$\bar{E} = (\bar{E} \bullet \bar{s})\bar{s}' + (\bar{E} \bullet \bar{t})\bar{t}' + (\bar{E} \bullet \bar{n})\bar{n}' \tag{5.46}$$

Alternatively, a vector can be represented in terms of its projections onto the reciprocal base vectors, or in terms of its *contravariant* components, leading to

$$\bar{E} = (\bar{E} \bullet \bar{s}')\bar{s} + (\bar{E} \bullet \bar{t}')\bar{t} + (\bar{E} \bullet \bar{n}')\bar{n} \tag{5.47}$$

The covariant components (such as $\bar{E} \bullet \bar{s}$) are the tangential components along the various curves defined by holding two of the three parameters (u, v, w) constant, while the contravariant components (such as $\bar{E} \bullet \bar{s}'$) are the components perpendicular to the constant parameter surfaces. Of course, (5.46) and (5.47) are just different ways of expressing the same vector function $\bar{E}$.

When discretizing vector integral equations, we employ either divergence-conforming or curl-conforming basis functions to represent the unknown vector quantities. With divergence-conforming functions, our goal is to maintain the normal-vector continuity of the function across cell boundaries, and the interpolation properties (if any) associated with "normal" components at various locations within the cell. Since the quantities of interest are the normal vector components, it will be convenient to work with the contravariant components as expressed in (5.47).

In contrast, when employing curl-conforming basis functions, our goal is usually to maintain the tangential-vector continuity between cells and to maintain the interpolation properties, if any, of the "tangential" components of those basis functions. Thus, the natural approach is to work with the covariant components of the functions as in (5.46).

5.4   RESTRICTION TO SURFACES

Our immediate problem of interest involves the mapping from a square or triangular planar reference cell (2D) to a curved surface in a 3D space. In the reference cell, the third variable (w) is not involved. The mapping functions x, y, and z are only functions of u and v. In common with our previous approach, the position vector from the origin to a point (x, y, z) is given by

$$\bar{r}(u, v) = x(u, v)\hat{x} + y(u, v)\hat{y} + z(u, v)\hat{z} \tag{5.48}$$

The base vectors for this cell are

$$\bar{s} = \frac{\partial \bar{r}}{\partial u} = \frac{\partial x}{\partial u}\hat{x} + \frac{\partial y}{\partial u}\hat{y} + \frac{\partial z}{\partial u}\hat{z} \tag{5.49}$$

$$\bar{t} = \frac{\partial \bar{r}}{\partial v} = \frac{\partial x}{\partial v}\hat{x} + \frac{\partial y}{\partial v}\hat{y} + \frac{\partial z}{\partial v}\hat{z} \tag{5.50}$$

while the reciprocal base vectors are

$$\bar{s}' = \nabla u = \frac{\partial u}{\partial x}\hat{x} + \frac{\partial u}{\partial y}\hat{y} + \frac{\partial u}{\partial z}\hat{z} \tag{5.51}$$

$$\bar{t}' = \nabla v = \frac{\partial v}{\partial x}\hat{x} + \frac{\partial v}{\partial y}\hat{y} + \frac{\partial v}{\partial z}\hat{z} \tag{5.52}$$

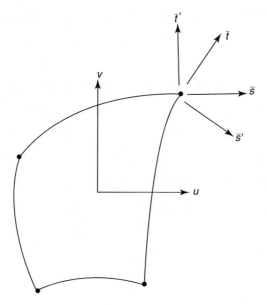

FIGURE 5.1: Base vectors $\bar{s}$, $\bar{t}$ and reciprocal base vectors $\bar{s}'$, $\bar{t}'$ on a curvilinear patch.

The base vectors are tangential to the surface defined by (5.48), while the reciprocal base vectors are normal to constant-u or constant-v surfaces, respectively. These vectors are illustrated in Fig. 5.1.

Derivatives transform according to

$$
\begin{bmatrix} \dfrac{\partial}{\partial u} \\[2ex] \dfrac{\partial}{\partial v} \end{bmatrix} = \begin{bmatrix} \dfrac{\partial x}{\partial u} & \dfrac{\partial y}{\partial u} & \dfrac{\partial z}{\partial u} \\[2ex] \dfrac{\partial x}{\partial v} & \dfrac{\partial y}{\partial v} & \dfrac{\partial z}{\partial v} \end{bmatrix} \begin{bmatrix} \dfrac{\partial}{\partial x} \\[2ex] \dfrac{\partial}{\partial y} \\[2ex] \dfrac{\partial}{\partial z} \end{bmatrix}
\tag{5.53}
$$

and

$$
\begin{bmatrix} \dfrac{\partial}{\partial x} \\[2ex] \dfrac{\partial}{\partial y} \\[2ex] \dfrac{\partial}{\partial z} \end{bmatrix} = \begin{bmatrix} \dfrac{\partial u}{\partial x} & \dfrac{\partial v}{\partial x} \\[2ex] \dfrac{\partial u}{\partial y} & \dfrac{\partial v}{\partial y} \\[2ex] \dfrac{\partial u}{\partial z} & \dfrac{\partial v}{\partial z} \end{bmatrix} \begin{bmatrix} \dfrac{\partial}{\partial u} \\[2ex] \dfrac{\partial}{\partial v} \end{bmatrix}
\tag{5.54}
$$

The Jacobian matrix in (5.53) has dimension 2×3 in accordance with the dimensionality of the mapping, while the inverse Jacobian matrix in (5.54) has dimension 3×2.

Since the functions x, y, and z are defined explicitly (Chapter 2), the entries of the Jacobian matrix in (5.53) are readily available. However, the matrix entries in (5.54) are not. In situations where the inverse Jacobian entries are required (to be encountered below), it is necessary to be able to invert the Jacobian matrix numerically, which is not possible if it is not a square matrix! Thus, it is desirable to introduce a dummy parameter w, to fill the equations in (5.53) and (5.54) out to 3×3 systems. Since the third variable is arbitrary, the dummy parameter may be constrained so that

$$\bar{n}' = \nabla w = \frac{\bar{s} \times \bar{t}}{|\bar{s} \times \bar{t}|} \tag{5.55}$$

which, by virtue of its definition, is a unit vector. (The only assumption being made here is that w can be chosen to make $\bar{n}'$ a unit vector.) From (5.33),

$$|\bar{s} \times \bar{t}| = \sqrt{\left(\frac{\partial y}{\partial u}\frac{\partial z}{\partial v} - \frac{\partial z}{\partial u}\frac{\partial y}{\partial v}\right)^2 + \left(\frac{\partial z}{\partial u}\frac{\partial x}{\partial v} - \frac{\partial x}{\partial u}\frac{\partial z}{\partial v}\right)^2 + \left(\frac{\partial x}{\partial u}\frac{\partial y}{\partial v} - \frac{\partial y}{\partial u}\frac{\partial x}{\partial v}\right)^2}$$
$$= D(u, v) \tag{5.56}$$

However, Eq. (5.39) still holds in this situation, implying that the quantity equivalent to the determinant in this case is given by

$$\det \mathbf{J} = D \tag{5.57}$$

From (5.35), it appears that one way by which (5.57) can be true is if

$$\bar{n} = \frac{\partial \bar{r}}{\partial w} = \frac{1}{D}\left(\frac{\partial y}{\partial u}\frac{\partial z}{\partial v} - \frac{\partial z}{\partial u}\frac{\partial y}{\partial v}\right)\hat{x} + \frac{1}{D}\left(\frac{\partial z}{\partial u}\frac{\partial x}{\partial v} - \frac{\partial x}{\partial u}\frac{\partial z}{\partial v}\right)\hat{y}$$
$$+ \frac{1}{D}\left(\frac{\partial x}{\partial u}\frac{\partial y}{\partial v} - \frac{\partial y}{\partial u}\frac{\partial x}{\partial v}\right)\hat{z} \tag{5.58}$$

which is the same as

$$\bar{n} = \frac{1}{D}\bar{s} \times \bar{t} = \bar{n}' \tag{5.59}$$

The implications of this choice for w are summarized as follows:

1. $\hat{n} = \hat{n}'$, and both are unit vectors normal to the surface of constant w.

2. Vectors $\bar{s}$ and $\bar{t}$ are tangential to the surface of constant w, by definition. Since $\bar{n}$ is perpendicular to the surface (by our choice of w), the reciprocal base vectors $\bar{s}'$ and $\bar{t}'$ are also tangential to the surface of constant w. This conclusion follows from (5.40) and (5.41), which in this case simplify to

$$\bar{s}' = \frac{1}{D}\bar{t} \times \bar{n} \tag{5.60}$$

$$\bar{t}' = \frac{1}{D}\bar{n} \times \bar{s} \tag{5.61}$$

3. Since (a) $\bar{n}$ is a unit vector, (b) $\bar{n}$ is perpendicular to $\bar{t}$, and (c) $\bar{n}$ is perpendicular to $\bar{s}$, the magnitudes of the reciprocal base vectors are related to those of the base vectors by

$$\left|\bar{s}'\right| = \left|\frac{1}{D}\bar{t} \times \hat{n}\right| = \frac{|\bar{t}|}{D} \tag{5.62}$$

$$\left|\bar{t}'\right| = \left|\frac{1}{D}\hat{n} \times \bar{s}\right| = \frac{|\bar{s}|}{D} \tag{5.63}$$

4. The explicit form of the 3×3 Jacobian matrix is given by

$$\mathbf{J} = \begin{bmatrix} \dfrac{\partial x}{\partial u} & \dfrac{\partial y}{\partial u} & \dfrac{\partial z}{\partial u} \\[2mm] \dfrac{\partial x}{\partial v} & \dfrac{\partial y}{\partial v} & \dfrac{\partial z}{\partial v} \\[2mm] \dfrac{1}{D}\left(\dfrac{\partial y}{\partial u}\dfrac{\partial z}{\partial v} - \dfrac{\partial z}{\partial u}\dfrac{\partial y}{\partial v}\right) & \dfrac{1}{D}\left(\dfrac{\partial z}{\partial u}\dfrac{\partial x}{\partial v} - \dfrac{\partial x}{\partial u}\dfrac{\partial z}{\partial v}\right) & \dfrac{1}{D}\left(\dfrac{\partial x}{\partial u}\dfrac{\partial y}{\partial v} - \dfrac{\partial y}{\partial u}\dfrac{\partial x}{\partial v}\right) \end{bmatrix} \tag{5.64}$$

and can be used to compute the inverse Jacobian matrix entries in (5.54).

In summary, when working with the base and reciprocal base vectors on a 2D surface in 3D space, it will be convenient to invoke the assumption in (5.55).

This ensures that base vectors $\bar{s}$ and $\bar{t}$, and reciprocal base vectors $\bar{s}'$ and $\bar{t}'$, are tangential to the surface and permits the computation of all needed parameters.

5.5 CURL-CONFORMING BASIS FUNCTIONS ON CURVILINEAR CELLS

We now turn our attention to the mapping of the vector basis functions from the reference cell in (u, v) space to the curvilinear cell in (x, y, z) space. It is simpler to discuss the mapping of curl-conforming basis functions, so we consider those first. Curl-conforming functions maintain tangential-vector continuity across cell boundaries, and our mapping procedure must ensure that behavior. Since the principal quantities of interest are the tangential vector components at cell boundaries, it is natural to work with the covariant components of the basis functions as expressed in (5.46). We desire to match the tangential components of the basis functions (where "tangential" corresponds to the directions of the base vectors $\bar{s}$ and $\bar{t}$, even at locations away from the cell boundary) on the curvilinear cell with those on the reference cell:

$$\bar{B} \bullet \bar{s} = R_u^{\text{curl}} \tag{5.65}$$

$$\bar{B} \bullet \bar{t} = R_v^{\text{curl}} \tag{5.66}$$

In accordance with (5.46), the basis function can be written as

$$\bar{B} = R_u^{\text{curl}} \bar{s}' + R_v^{\text{curl}} \bar{t}' \tag{5.67}$$

On a component-by-component level, (5.67) is equivalent to the matrix relation

$$\begin{bmatrix} B_x \\ B_y \\ B_z \end{bmatrix} = \begin{bmatrix} \dfrac{\partial u}{\partial x} & \dfrac{\partial v}{\partial x} \\ \dfrac{\partial u}{\partial y} & \dfrac{\partial v}{\partial y} \\ \dfrac{\partial u}{\partial z} & \dfrac{\partial v}{\partial z} \end{bmatrix} \begin{bmatrix} R_u^{\text{curl}} \\ R_v^{\text{curl}} \end{bmatrix} = \mathbf{J}^{-1} \begin{bmatrix} R_u^{\text{curl}} \\ R_v^{\text{curl}} \end{bmatrix} \tag{5.68}$$

where $\mathbf{J}^{-1}$ is used symbolically to denote the 3×2 matrix in (5.54) and (5.68) and not necessarily the "inverse" of $\mathbf{J}$, which must be obtained from the 3×3 system in

(5.64). There are two aspects of (5.68) to consider, (1) the continuity of tangential fields across cell boundaries, and (2) the normalization of the basis functions in the curvilinear domain.

As a consequence of the way that the mapping of the cell coordinates is defined (Chapter 2), the vectors $\bar{s}$ and $\bar{t}$ are tangential to the boundaries of the curvilinear patch at the extreme values of the parameters v and u. Within two adjacent cells sharing a boundary that coincides with a constant-u surface, and common cell endpoints in v, the base vector tangential to that boundary is defined by the derivatives of x, y, and z with respect to v, and will therefore be the same in either cell. The imposition of (5.65) and (5.66) in adjacent cells, for basis functions that are adjusted to maintain tangential continuity in the reference space, will ensure that the resulting vector basis functions in (x, y, z) space also maintain tangential continuity.

The curl-conforming basis functions (Chapter 4) can be normalized so that their tangential components have unity value at appropriate locations along a cell boundary. Consider one such location v_i along a boundary where $\bar{t}$ is tangential and R_v^{curl} has a unit tangent. In the curvilinear space, it follows that

$$\bar{B} \bullet \bar{t}\Big|_{v=v_i} = 1 \qquad (5.69)$$

Since the vector $\bar{t}$ is not a unit vector, (5.69) implies that the tangential component in the curvilinear space is

$$\bar{B} \bullet \hat{t}\Big|_{v=v_i} = \frac{1}{|\bar{t}|}\bigg|_{v=v_i} \qquad (5.70)$$

where $\hat{t}$ is a unit vector in the $\bar{t}$ direction. Consequently, an additional scaling is necessary before the curl-conforming basis functions in the curvilinear space exhibit unity tangential components. In a similar manner, at a location along a boundary where $\bar{s}$ is tangential and R_u^{curl} has a unity tangential component,

$$\bar{B} \bullet \hat{s}\Big|_{u=u_i} = \frac{1}{|\bar{s}|}\bigg|_{u=u_i} \qquad (5.71)$$

These scale factors are constant within a cell, since they are base vectors evaluated at specific locations, but they differ with the basis function and from cell to cell in general. If it is necessary to compute the value of $\bar{B}$ at a point within the cell, the 3×3 matrix in (5.64) may be evaluated at that point and inverted to provide numerical values for the reciprocal base vector components.

5.6   DIVERGENCE-CONFORMING BASIS FUNCTIONS ON CURVILINEAR CELLS

Divergence-conforming functions maintain normal-vector continuity across cell boundaries, and our mapping procedure must ensure that behavior in x–y–z space. Since the principal quantities of interest are the normal-vector components on cell boundaries, it will be necessary to work with the contravariant components of the basis functions as expressed in (5.47).

While the mapping of the cell coordinates as defined in Chapter 2 ensures that the tangential base vectors on either side of a common cell boundary are the same, it does not guarantee that property for the reciprocal base vectors normal to a common boundary. For a boundary defined by a constant value of parameter u, for instance, the base vector $\bar{t}$ (tangential to that boundary) is a sole function of v, which has the same value at any location along the boundary for both cells sharing that boundary. The reciprocal base vector $\bar{s}'$ (normal to that same boundary) is also a function of u, meaning that it generally will not be the same on either side of a constant-u boundary.

The magnitude of $\bar{s}'$ at any location within a cell was determined in (5.62) to be

$$|\bar{s}'| = \frac{|\bar{t}|}{D} \tag{5.72}$$

where D is defined in (5.56). Since this general expression is the same in either cell sharing a constant-u boundary, and $\bar{t}$ is the same on either side of that same cell boundary, it is sufficient to scale by the function D to obtain vectors of the same length along the boundary in either cell. Similarly, on a boundary defined by a constant value for v, the magnitude of $\vec{t}\,'$ is

$$|\vec{t}\,'| = \frac{|\bar{s}|}{D} \tag{5.73}$$

Since $\bar{s}$ is the same on either side of a constant-v boundary, it is again sufficient to scale by D. (Note that D is generally not a constant along the boundary or within the cell.)

Thus, to impose normal-vector continuity, we will need to equate the contravariant components of the basis functions in (x, y, z) space with the corresponding components of divergence-conforming functions on the reference cell, scaled by the function $D(u, v)$:

$$\bar{B} \bullet \bar{s}\,' = \frac{1}{D} R_u^{\mathrm{div}} \tag{5.74}$$

$$\bar{B} \bullet \vec{t}\,' = \frac{1}{D} R_v^{\mathrm{div}} \tag{5.75}$$

Equivalently, in accordance with (5.47), the basis function can be written as

$$\bar{B} = \frac{1}{D} R_u^{\mathrm{div}} \bar{s} + \frac{1}{D} R_v^{\mathrm{div}} \vec{t} \tag{5.76}$$

Equation (5.76) is the same as the matrix relation

$$\begin{bmatrix} B_x \\ B_y \\ B_z \end{bmatrix} = \frac{1}{D} \begin{bmatrix} \dfrac{\partial x}{\partial u} & \dfrac{\partial x}{\partial v} \\ \dfrac{\partial y}{\partial u} & \dfrac{\partial y}{\partial v} \\ \dfrac{\partial z}{\partial u} & \dfrac{\partial z}{\partial v} \end{bmatrix} \begin{bmatrix} R_u^{\mathrm{div}} \\ R_v^{\mathrm{div}} \end{bmatrix} = \frac{1}{D} \mathbf{J}^{\mathrm{T}} \begin{bmatrix} R_u^{\mathrm{div}} \\ R_v^{\mathrm{div}} \end{bmatrix} \tag{5.77}$$

where $\mathbf{J}^{\mathrm{T}}$ is the transpose of the matrix in (5.53).

The divergence-conforming basis functions presented in Chapter 3 are normalized so that their normal components have unity value at appropriate locations along a cell boundary. From the definition in (5.77), at some location v_i along a constant-u boundary where $\bar{s}\,'$ is normal and R_u^{div} has a unit normal component, the basis function in the curvilinear space satisfies

$$\bar{B} \bullet \bar{s}\,' \big|_{v=v_i} = \frac{1}{D} \tag{5.78}$$

Because $\bar{s}'$ is not a unit vector, (5.78) implies that the normal component of the basis function is

$$\bar{B} \bullet \hat{s}'\Big|_{v=v_i} = \frac{1}{D|\bar{s}'|}\bigg|_{v=v_i} \tag{5.79}$$

Since the reciprocal base vectors are not explicitly defined by the mapping, it is convenient to employ (5.62) to obtain the equivalent result

$$\bar{B} \bullet \hat{s}'\Big|_{v=v_i} = \frac{1}{|\bar{t}|}\bigg|_{v=v_i} \tag{5.80}$$

Therefore, this basis function must be scaled by the magnitude of $\bar{t}$ at the interpolation point v_i so that its normal component exhibits a unity value there. Similarly, at a location along a boundary where $\bar{t}'$ is normal and R_v^{div} has a unity normal component, analogous reasoning leads to the result

$$\bar{B} \bullet \hat{t}'\Big|_{u=u_i} = \frac{1}{|\bar{s}|}\bigg|_{u=u_i} \tag{5.81}$$

These scale factors are constant within a cell, but may be different for each basis function and each cell.

Since the mapping (Chapter 2) provides an explicit expression for x, y, z and their derivatives, the basis functions are easily computed from (5.77) at points within the curvilinear cell.

5.7   THE IMPLEMENTATION OF VECTOR DERIVATIVES

An intrinsic feature of the type of analysis under consideration is that all operations involving the basis functions on the curvilinear cell can be transferred to the reference cell. First, we note that the surface gradient operator may be expressed as

$$\begin{aligned}
\nabla f &= \frac{\partial f}{\partial u}\nabla u + \frac{\partial f}{\partial v}\nabla v \\
&= \frac{\partial f}{\partial u}\bar{s}' + \frac{\partial f}{\partial v}\bar{t}'
\end{aligned} \tag{5.82}$$

Consider a curl-conforming basis function defined by

$$\begin{aligned}
\bar{B} &= R_u^{\text{curl}} \bar{s}' + R_v^{\text{curl}} \bar{t}' \\
&= R_u^{\text{curl}} \nabla u + R_v^{\text{curl}} \nabla v
\end{aligned} \tag{5.83}$$

Using the standard vector identity

$$\nabla \times (f \nabla g) = \nabla f \times \nabla g \tag{5.84}$$

the curl of $\bar{B}$ can be expanded to produce

$$\begin{aligned}
\nabla \times \bar{B} &= \nabla \times \left(R_u^{\text{curl}} \nabla u \right) + \nabla \times \left(R_v^{\text{curl}} \nabla v \right) \\
&= \nabla \left(R_u^{\text{curl}} \right) \times \nabla u + \nabla \left(R_v^{\text{curl}} \right) \times \nabla v \\
&= \nabla \left(R_u^{\text{curl}} \right) \times \bar{s}' + \nabla \left(R_v^{\text{curl}} \right) \times \bar{t}'
\end{aligned} \tag{5.85}$$

From (5.82)

$$\nabla \left(R_u^{\text{curl}} \right) = \frac{\partial R_u^{\text{curl}}}{\partial u} \bar{s}' + \frac{\partial R_u^{\text{curl}}}{\partial v} \bar{t}' \tag{5.86}$$

$$\nabla \left(R_v^{\text{curl}} \right) = \frac{\partial R_v^{\text{curl}}}{\partial u} \bar{s}' + \frac{\partial R_v^{\text{curl}}}{\partial v} \bar{t}' \tag{5.87}$$

Thus, Eq. (5.85) can be written as

$$\nabla \times \bar{B} = \frac{\partial R_u^{\text{curl}}}{\partial u} \bar{s}' \times \bar{s}' + \frac{\partial R_u^{\text{curl}}}{\partial v} \bar{t}' \times \bar{s}' + \frac{\partial R_v^{\text{curl}}}{\partial u} \bar{s}' \times \bar{t}' + \frac{\partial R_v^{\text{curl}}}{\partial v} \bar{t}' \times \bar{t}' \tag{5.88}$$

The first and last terms on the right side of (5.88) vanish, leaving

$$\begin{aligned}
\nabla \times \bar{B} &= \frac{\partial R_u^{\text{curl}}}{\partial v} \bar{t}' \times \bar{s}' + \frac{\partial R_v^{\text{curl}}}{\partial u} \bar{s}' \times \bar{t}' \\
&= \left\{ \frac{\partial R_v^{\text{curl}}}{\partial u} - \frac{\partial R_u^{\text{curl}}}{\partial v} \right\} (\bar{s}' \times \bar{t}')
\end{aligned} \tag{5.89}$$

Finally, using (5.45), we obtain

$$\nabla \times \bar{B} = \frac{1}{D} \left\{ \frac{\partial R_v^{\text{curl}}}{\partial u} - \frac{\partial R_u^{\text{curl}}}{\partial v} \right\} \bar{n} \tag{5.90}$$

where $\bar{n}$ is a unit vector in accordance with the development in Section 5.4. The curl of $\bar{B}$ in x–y–z space may be obtained directly in terms of the reference coordinates u and v, using an expression (5.90) that is in fact a local curl operation scaled by D.

Consider a divergence-conforming basis function defined by

$$\bar{B} = \frac{1}{D} R_u^{\mathrm{div}} \bar{s} + \frac{1}{D} R_v^{\mathrm{div}} \bar{t}$$
$$= R_u^{\mathrm{div}}(\bar{t}' \times \bar{n}') + R_v^{\mathrm{div}}(\bar{n}' \times \bar{s}') \tag{5.91}$$

Since $\bar{t}' \times \bar{n}' = \nabla v \times \nabla w$, the vector identities

$$\nabla \bullet (\bar{A} \times \bar{B}) = \bar{B} \bullet \nabla \times \bar{A} - \bar{A} \bullet \nabla \times \bar{B} \tag{5.92}$$

and

$$\nabla \times \nabla f = 0 \tag{5.93}$$

can be used to show that

$$\nabla \bullet (\bar{t}' \times \bar{n}') = 0 \tag{5.94}$$

It follows from the additional vector identity

$$\nabla \bullet (f\bar{V}) = f \nabla \bullet \bar{V} + \nabla f \bullet \bar{V} \tag{5.95}$$

that the divergence of $\bar{B}$ is given by

$$
\begin{aligned}
\nabla \bullet \bar{B} &= \nabla\left(R_u^{\mathrm{div}}\right) \bullet (\bar{t}' \times \bar{n}') + \nabla\left(R_v^{\mathrm{div}}\right) \bullet (\bar{n}' \times \bar{s}') \\
&= \nabla\left(R_u^{\mathrm{div}}\right) \bullet \frac{1}{D}\bar{s} + \nabla\left(R_v^{\mathrm{div}}\right) \bullet \frac{1}{D}\bar{t} \\
&= \frac{1}{D}\left\{ \frac{\partial R_u^{\mathrm{div}}}{\partial u}\bar{s}' + \frac{\partial R_u^{\mathrm{div}}}{\partial v}\bar{t}' \right\} \bullet \bar{s} + \left\{ \frac{\partial R_v^{\mathrm{div}}}{\partial u}\bar{s}' + \frac{\partial R_v^{\mathrm{div}}}{\partial v}\bar{t}' \right\} \bullet \bar{t} \\
&= \frac{1}{D}\left\{ \frac{\partial R_u^{\mathrm{div}}}{\partial u} + \frac{\partial R_v^{\mathrm{div}}}{\partial v} \right\}
\end{aligned}
\tag{5.96}
$$

Therefore, the divergence of (5.91) in x–y–z space can be calculated directly from the application of a divergence operator in the reference coordinates, scaled by the function D.

5.8 SUMMARY

This chapter has presented a procedure for mapping divergence-conforming and curl-conforming basis functions to a curvilinear surface in 3D space. This process ensures the continuity of normal components at cell boundaries (divergence-conforming) or tangential components at cell boundaries (curl-conforming) as required. The normalization of the basis functions is considered, along with a procedure for transferring vector derivatives from the curvilinear coordinates to the reference coordinates. These results will be used in Chapters 6 and 7.

REFERENCES

[1] J. A. Stratton, *Electromagnetic Theory*. New York: McGraw-Hill, 1941.

[2] C. W. Crowley, "Mixed-order Covariant Projection Finite Elements for Vector Fields," Ph.D. Dissertation, McGill University, Montreal, Quebec, 1988.

[3] S. Wandzura, "Electric current basis functions for curved surfaces," *Electromagnetics*, vol. 12, pp. 77–91, 1992.

[4] G. E. Antilla and N. G. Alexopoulos, "Scattering from complex three-dimensional geometries by a curvilinear hybrid finite element-integral equation approach," *J. Opt. Soc. Am. A*, vol. 11, pp. 1445–1457, 1994.

[5] A. F. Peterson and K. R. Aberegg, "Parametric mapping of vector basis functions for surface integral equation formulations," *ACES J.*, vol. 10, pp. 107–115, 1995.

CHAPTER 6

Use of Divergence-conforming Basis Functions with the Electric Field Integral Equation

To demonstrate the vector mapping procedures developed in Chapter 5, we consider the electric field integral equation (EFIE) for scattering from perfectly conducting objects with curved surfaces. The EFIE was derived in Chapter 1. Since the EFIE operator involves a divergence of the surface current density, the method of moments (MoM) procedure will be illustrated in conjunction with divergence-conforming basis and testing functions. Expressions for the MoM matrix entries incorporating the curved-cell mappings are developed. Several numerical results are presented for illustration.

6.1 TESTED FORM OF THE EFIE

The MoM procedure employs a testing function to convert the equation under consideration into a weak form, which reduces the number of derivatives acting on the unknown function. If the EFIE of Eq. (1.8) is multiplied (by scalar product) with a vector testing function $\bar{T}(s, t)$, tangential to the surface, one obtains the equation

$$\iint_{\text{surface}} \bar{T} \bullet \bar{E}^{\text{inc}} ds \, dt = - \iint_{\text{surface}} \bar{T} \bullet \bar{E}^s ds \, dt \qquad (6.1)$$

where the integration is performed over the surface of the conductor. The incident electric field $\bar{E}^{\text{inc}}$ is the field in the absence of the structure (the excitation) and is assumed known. The function $\bar{E}^s$ is the "scattered" electric field produced by the surface current $\bar{J}$, also in the absence of the structure, and may be obtained from the expression

$$\bar{E}^s = -j\omega\mu \iint_{\text{surface}} \bar{J}(s', t')G \, ds' \, dt' + \frac{1}{j\omega\varepsilon} \nabla \iint_{\text{surface}} \nabla' \bullet \bar{J}(s', t')G \, ds' \, dt' \qquad (6.2)$$

where G is the free space Green's function

$$G(R) = \frac{e^{-jkR}}{4\pi R} \qquad (6.3)$$

and

$$R = \sqrt{[x(s, t) - x(s', t')]^2 + [y(s, t) - y(s', t')]^2 + [z(s, t) - z(s', t')]^2} \qquad (6.4)$$

To obtain (6.2) from Eq. (1.8), we employed the well-known relation

$$\nabla \bullet \bar{A} = \iint_{\text{surface}} \nabla' \bullet \bar{J}(s', t')G(R)ds' \, dt' \qquad (6.5)$$

The operator in (6.2) involves a divergence of the surface current density, and provides motivation for the use of divergence-conforming basis functions.

6.2   THE SUBSECTIONAL MODEL

The surface of the conducting object is to be represented by curved cells having either three sides or four sides. Such a mapping may be obtained using the quadratic Lagrangian interpolation polynomials of Sections 2.3 and 2.5, and reference cells that are triangular or square. The cell shapes are defined by explicit functions of local parameters u and v, including Eqs. (2.29) and (2.45) for x, and equivalent expressions for y and z. Since these equations are to be incorporated into the computer program that performs the analysis, the curved-cell model that must be provided as input to the analysis program consists of a list of the coordinates of the nodes that define each cell (nine nodes for square cells, six for triangular) and connectivity arrays that link the cells, edges, and nodes on the surface.

The surface current density in (6.2) is represented by vector basis functions $\{\bar{B}_n\}$ that are distributed over the curved-cell surface model. As discussed in Chapter 3, the divergence-conforming basis functions either straddle two adjacent cells or are entirely confined within a single cell. The connectivity arrays associated with the surface can be used to provide a systematic way of organizing the basis functions within a cell and linking edge-based basis functions that straddle cells to appropriate basis functions in adjacent cells. The current is represented by the summation

$$\bar{J}(s,\ t) = \sum_{n=1}^{N} I_n \bar{B}_n(s,\ t) \tag{6.6}$$

where $\{I_n\}$ denote N complex-valued coefficients that henceforth are the unknowns to be determined. The discretized EFIE exhibits symmetry between the basis functions and testing functions—look ahead to Eq. (6.13)—and it is therefore convenient to define the testing functions to be the same as the divergence-conforming basis functions

$$\bar{T}_m(s,\ t) = \bar{B}_m(s,\ t) \tag{6.7}$$

When used within (6.1), the testing functions provide a means for obtaining N linearly independent equations from the EFIE. The equations can be organized into the form of a matrix equation

$$\mathbf{E} = \mathbf{ZI} \tag{6.8}$$

where $\mathbf{E}$ and $\mathbf{I}$ are $N \times 1$ column vectors and $\mathbf{Z}$ is an $N \times N$ matrix. The entries of $\mathbf{I}$ are the coefficients $\{I_n\}$, while those of $\mathbf{E}$ are given by

$$E_m = \iint\limits_{\text{surface}} \bar{T}_m \bullet \bar{E}^{\text{inc}} ds \, dt \tag{6.9}$$

The entries of $\mathbf{Z}$ have the general form

$$Z_{mn} = - \iint\limits_{\text{surface}} \bar{T}_m \bullet \bar{E}_n^{\text{s}} ds \, dt \tag{6.10}$$

where $\bar{E}_n^{\text{s}}$ denotes the electric field produced by the nth basis function. Although this quantity is defined by the expression

$$\bar{E}_n^{\text{s}} = -j\omega\mu \iint\limits_{\text{surface}} \bar{B}_n(s', t')G \, ds'dt' + \frac{1}{j\omega\varepsilon}\nabla \iint\limits_{\text{surface}} \nabla' \bullet \bar{B}_n(s', t')G \, ds'dt' \tag{6.11}$$

the form of (6.11) does not exploit the testing function to eliminate a derivative. Instead, using the vector identity

$$\nabla \bullet (f \, \bar{T}) = \bar{T} \bullet \nabla f + f \, \nabla \bullet \bar{T} \tag{6.12}$$

in conjunction with the divergence theorem, (6.10) may be recast as

$$\begin{aligned} Z_{mn} = j\omega\mu \iint\limits_{s,t} \iint\limits_{s',t'} \bar{T}_m(s,t) \bullet \bar{B}_n(s',t')G \, ds' \, dt' \, ds \, dt \\ + \frac{1}{j\omega\varepsilon} \iint\limits_{s,t} \nabla \bullet \bar{T}_m \iint\limits_{s',t'} \nabla' \bullet \bar{B}_n G \, ds'dt'ds \, dt \end{aligned} \tag{6.13}$$

The matrix equation in (6.8) may be solved to produce the coefficients $\{I_n\}$, after which any other quantity of interest may be obtained by integrating over the current density in (6.6).

If (6.13) is used with a basis function that is not divergence-conforming, the divergence operation would produce a Dirac delta function type of behavior at any locations where the function lacks normal-vector continuity. In special cases when it might be necessary to employ nondivergence-conforming basis functions (such as to incorporate lumped-element feeds or loads, junctions between surfaces and wires, etc.) the Dirac delta function must be included as part of the expression. The result is equivalent to including a line integral along with the surface integrals in (6.13).

6.3   MAPPED MoM MATRIX ENTRIES

Since each cell comprising the surface is defined by an independent mapping, the preceding integral expressions must be evaluated separately for each cell. The *observer* cells where the testing functions reside are generally different from the *source* cells where the basis functions reside. A single matrix entry may require integrals over as many as four cells, since each edge-based basis or testing function straddles one or two cells. In the following, an index (m or n) is used to denote specific testing and basis functions, and the discussion considers a single observer cell and a single source cell. The pointer arrays associated with the surface model are used to identify the appropriate cells where these functions reside.

The integrals required in (6.9) and (6.13) must be transformed into integrals in the local coordinates of each reference cell, as explained in Chapter 5. The divergence-conforming basis functions are defined by the transformation in Eq. (5.77):

$$
\begin{bmatrix} B_x \\ B_y \\ B_z \end{bmatrix} = \frac{1}{D} \begin{bmatrix} \dfrac{\partial x}{\partial u} & \dfrac{\partial x}{\partial v} \\[2mm] \dfrac{\partial y}{\partial u} & \dfrac{\partial y}{\partial v} \\[2mm] \dfrac{\partial z}{\partial u} & \dfrac{\partial z}{\partial v} \end{bmatrix} \begin{bmatrix} R_u^{\mathrm{div}} \\ R_v^{\mathrm{div}} \end{bmatrix} = \frac{1}{D} \mathbf{J}^{\mathrm{T}} \begin{bmatrix} R_u^{\mathrm{div}} \\ R_v^{\mathrm{div}} \end{bmatrix} \tag{6.14}
$$

where $\mathbf{J}^{\mathrm{T}}$ is used to denote the transpose of the Jacobian matrix in (5.53), and $\bar{R}^{\mathrm{div}}$ is the basis function in the reference cell. (The basis functions may be normalized so that desired components have unity value at certain locations on the curved cell; for the present discussion we omit the normalization constant.) The testing functions are also to be divergence-conforming functions, defined by a mapping of the same form. (The normalization of the testing functions, as described in Section 5.6, is optional.)

The evaluation of the preceding integrals proceeds with the assistance of matrix relations such as

$$\begin{bmatrix} T_x^{\text{div}} \\ T_y^{\text{div}} \\ T_z^{\text{div}} \end{bmatrix}^{\text{T}} = \frac{1}{D(s,t)|_{\text{observer}}} \begin{bmatrix} R_u^{\text{div}} & R_v^{\text{div}} \end{bmatrix} \mathbf{J}|_{\text{observer}} \tag{6.15}$$

For instance, the entries of the excitation column vector $\mathbf{E}$ in (6.9) may be obtained as

$$\iint\limits_{\text{surface}} \bar{T} \bullet \bar{E}^{\text{inc}} \, ds \, dt = \iint\limits_{\text{reference cell (observer)}} \begin{bmatrix} R_u^{\text{div}} & R_v^{\text{div}} \end{bmatrix}_m \mathbf{J}|_{\text{observer}} \begin{bmatrix} E_x^{\text{inc}} \\ E_y^{\text{inc}} \\ E_z^{\text{inc}} \end{bmatrix} du \, dv \tag{6.16}$$

Note that the scale factor $D(s,t)$ in (6.15) is cancelled by the same factor within the differential surface area

$$ds \, dt = D(s,t)|_{\text{observer}} du \, dv \tag{6.17}$$

Standard matrix multiplication is used to collect terms in the integrand of (6.16), reducing the chain of three matrices to a single scalar quantity.

A similar approach can be used to express the dot product in the first integral of (6.13) as

$$\bar{T}_m \bullet \bar{B}_n = \begin{bmatrix} T_{mx}^{\text{div}} & T_{my}^{\text{div}} & T_{mz}^{\text{div}} \end{bmatrix} \begin{bmatrix} B_{nx}^{\text{div}} \\ B_{ny}^{\text{div}} \\ B_{nz}^{\text{div}} \end{bmatrix}$$

$$= \frac{1}{D(s,t)|_{\text{observer}}} \frac{1}{D(s',t')|_{\text{source}}} \begin{bmatrix} R_u^{\text{div}} & R_v^{\text{div}} \end{bmatrix}_m \mathbf{J}|_{\text{observer}} \mathbf{J}^{\text{T}}|_{\text{source}} \begin{bmatrix} R_u^{\text{div}} \\ R_v^{\text{div}} \end{bmatrix}_n \tag{6.18}$$

In (6.18), it is explicit that the testing function is located in the observer cell, where the Jacobian matrix is $\mathbf{J}|_{\text{observer}}$, while the basis function resides in the source cell, where the matrix is $\mathbf{J}|_{\text{source}}$ (and the quantities are defined in terms of primed coordinates). The differential surface area for the source cell is

$$ds' dt' = D(s',t')|_{\text{source}} du' dv' \tag{6.19}$$

The index of the testing function, m, is also independent of the index of the basis function, n. (Depending on the context, these indices may refer to global numbering system throughout the entire surface or a local numbering system within the individual cell.)

Using (6.18), the first integral in (6.13) may be written in the reference cell coordinates as

$$\iint\limits_{s,\,t}\iint\limits_{s',\,t'} \bar{T}_m(s,\,t) \bullet \bar{B}_n(s',\,t') G\,ds'\,dt'\,ds\,dt$$

$$= \iint\limits_{\substack{\text{reference cell}\\ \text{(observer)}}} \iint\limits_{\substack{\text{reference cell}\\ \text{(source)}}} \left[R_u^{\text{div}}\ R_v^{\text{div}} \right]_m \mathbf{J}|_{\text{observer}}\,\mathbf{J}^{\mathbf{T}}|_{\text{source}} \begin{bmatrix} R_u^{\text{div}} \\ R_v^{\text{div}} \end{bmatrix}_n G\,du'\,dv'\,du\,dv$$

$$(6.20)$$

In the integrand of (6.20), the scale factors $D(s,\,t)|_{\text{observer}}$ and $D(s',\,t')|_{\text{source}}$ in (6.18) cancel with those in (6.17) and (6.19). This integral must be evaluated by numerical quadrature carried out in the reference coordinates. The basis and testing functions, and the mapping functions $x(u,\,v)$, $y(u,\,v)$, and $z(u,\,v)$, are computed at the quadrature points for both the source and observer reference cells. The entries of the two Jacobian matrices in (6.20) are also required at the quadrature points; these are easily obtained from (5.53) and the explicit expressions for the derivatives of x, y, and z.

The second integral in (6.13) involves the divergence of the testing function and basis function. These can be obtained using (5.96), which is equivalent to

$$\nabla \bullet \bar{T}_m = \frac{1}{D(s,\,t)|_{\text{observer}}} \left\{ \frac{\partial R_u^{\text{div}}}{\partial u} + \frac{\partial R_v^{\text{div}}}{\partial v} \right\}_m \tag{6.21}$$

$$\nabla' \bullet \bar{B}_n = \frac{1}{D(s',\,t')|_{\text{source}}} \left\{ \frac{\partial R_u^{\text{div}}}{\partial u'} + \frac{\partial R_v^{\text{div}}}{\partial v'} \right\}_n \tag{6.22}$$

Therefore, the second integral in (6.13) can be written as

$$\iint\limits_{s,\,t} \nabla \bullet \bar{T}_m \iint\limits_{s',\,t'} \nabla' \bullet \bar{B}_n G\,ds'\,dt'\,ds\,dt$$

$$= \iint\limits_{\text{reference cell (observer)}} \left\{ \frac{\partial R_u^{\text{div}}}{\partial u} + \frac{\partial R_v^{\text{div}}}{\partial v} \right\}_m \iint\limits_{\text{reference cell (source)}} \left\{ \frac{\partial R_u^{\text{div}}}{\partial u'} + \frac{\partial R_v^{\text{div}}}{\partial v'} \right\}_n G\,du'\,dv'\,du\,dv$$

$$(6.23)$$

As in (6.20), the scale factors $D(s,\,t)|_{\text{observer}}$ and $D(s',\,t')|_{\text{source}}$ are cancelled.

6.4   NORMALIZATION OF DIVERGENCE-CONFORMING BASIS FUNCTIONS

If it is desired to interpret the coefficients $\{I_n\}$ in Eq. (6.6) as values of the surface current density (interpolatory basis functions), the divergence-conforming basis functions used for $\bar{J}$ must be normalized so that the normal component of each $\bar{B}_n$ has unity value at an appropriate location within each cell (in the curvilinear x–y–z space). This is accomplished by scaling each basis function by a constant equal to the magnitude of the appropriate base vector at that location, as explained in Section 5.6. For functions that straddle two cells, the normalization constant may also contain a sign that ensures that the basis function points in a consistent normal-vector direction from a global perspective. The normalization is incorporated into the matrix entries of the preceding section by multiplying the basis functions in the reference cell by the appropriate constants. These constants are cell-specific, and can be determined at the start of the analysis and stored in an array for convenient reference as required during the matrix construction phase of the MoM procedure, and any post-processing that involves the basis functions.

6.5   TREATMENT OF THE SINGULARITY OF THE GREEN'S FUNCTION

When the source and observer cells in (6.20) and (6.23) coincide, there will generally be points where the function R within G vanishes. The resulting $1/R$ singularity complicates the evaluation of the integrals by quadrature. There are several possible approaches to evaluating the singular integral; here we consider the use of a Duffy transformation [1].

When the source and observer regions coincide, for any quadrature point arising in the evaluation of the *outer* integral, the domain of the inner integral in either (6.20) or (6.23) can be divided into three triangular subcells (if the reference cell is a triangle) or four triangular subcells (if the reference cell is square) with the singularity at one corner of each subcell. This approach reduces the integration to one of the form

$$I = \int\limits_{v=0}^{1} \int\limits_{u=0}^{v-1} f(u, v) \frac{1}{\sqrt{u^2 + (1-v)^2}} du\, dv \qquad (6.24)$$

where the function f incorporates the basis and testing functions, the numerator of the Green's function, the matrix products in (6.20), etc. The Duffy transformation involves a change of variable from u to w, where

$$u = (1 - v)w \qquad (6.25)$$

It follows from (6.25) that

$$du = (1 - v)dw \qquad (6.26)$$

The limits of integration are modified as well, with the lower limit $u = 0$ replaced by a new limit of $w = 0$, and the upper limit of $u = 1 - v$ replaced by $w = 1$. The transformation produces

$$\begin{aligned} I &= \int\limits_{v=0}^{1} \int\limits_{w=0}^{1} f(u, v) \frac{1-v}{\sqrt{u^2 + (1-v)^2}} dw\, dv \\ &= \int\limits_{v=0}^{1} \int\limits_{w=0}^{1} f(u, v) \frac{1}{\sqrt{w^2 + 1}} dw\, dv \end{aligned} \qquad (6.27)$$

The integrand in (6.27) is bounded over the entire domain. The transformation converts the triangular domain of (6.24) into a square domain, and in essence spreads the singularity over the additional edge to eliminate it. The evaluation by quadrature is straightforward, with the primary difference from the nonsingular case being that each inner integral is evaluated as three to four separate integrals.

The Duffy transformation may lead to irregular integrands in certain situations, such as the case when the aspect ratio of the subdivided cells is extreme (the observation point is very close to the original cell boundary). The matrix entries may also be difficult to evaluate when the observation point is outside the source cell but close to it. For these reasons, it is generally prudent to employ adaptive quadrature algorithms that can estimate the error in the integrations, seek a pre-scribed error level, and inform the user when they fail! Alternative approaches have been proposed that may improve upon the Duffy transformation [2, 3].

6.6   QUADRATURE RULES

Algorithms for numerical quadrature continue to evolve. The baseline approach for multidimensional domains is to implement Gauss–Legendre rules in product form. An adaptive alternative may be obtained using the Gauss–Kronrod–Patterson rules, which provide complete sample point reuse [4]. However, to treat 2D domains, it may be more efficient to employ rules that are specifically designed for square or triangular cells. A website with links to a large number of available rules has been developed by Cools [5].

6.7   EXAMPLE: SCATTERING CROSS SECTION OF A SPHERE

A computer program was constructed that implements the discretization of the EFIE using the $p = 0$ divergence-conforming functions of Chapter 3 with the quadratic cell shapes described in Sections 2.3 and 2.5. Table 6.1 shows some results obtained from the program, for flat and curved triangular-cell models of a sphere of radius 0.5λ, where λ is the free-space wavelength. It is convenient to consider a sphere since exact solutions are available for comparison. It is also convenient to examine the bistatic scattering cross section, which is defined in [6], since that parameter is a composite that depends on the currents produced on the entire sphere in response to a plane wave excitation. For the purpose of the comparisons, several models were created that employed flat triangular cells, and compared to similar models where the cells were curved to conform to the spherical surface. These models were constructed by dividing the sphere uniformly along θ, and subdividing in ϕ so that along the equator the triangle sides have the same dimension as they do in θ. In all cases, the surface area of the models is scaled to the same surface area as the desired sphere.

Table 6.1 presents the scattering cross section as a function of θ, for ϕ fixed at 0, in response to a plane wave propagating in the $\theta = 0$, $\phi = 0$ direction with the electric field polarized in the $\hat{x}$ direction. The table suggests that the curved-cell results converge to the exact solution as the model is refined. Furthermore, these results indicate that the 300-edge flat-cell model yields approximately the same accuracy as the 108-edge curved-cell model. The 300-edge model exhibits an average density of 96 unknowns/λ^2, roughly in the range where reasonably good solutions are expected from this type of formulation (at least for geometries as simple

TABLE 6.1: Scattering Cross Section Results

	FLAT		CURVED				EXACT
EDGES	192	300	48	108	192	300	
θ							
0	9.38	9.50	8.32	9.49	9.59	9.62	9.66
30	6.59	6.69	5.49	6.61	6.75	6.79	6.83
60	4.21	4.19	4.04	4.19	4.16	4.16	4.15
90	−6.45	−6.52	−5.37	−6.67	−6.60	−6.60	−6.58
120	1.65	1.64	1.13	1.88	1.67	1.64	1.63
150	−1.95	−1.77	−2.03	−1.28	−1.33	−1.37	−1.42
180	−2.86	−2.62	0.12	−2.60	−2.35	−2.30	−2.26

Note. Results given in dBλ^2 for sphere radius of 0.5λ, for flat triangular-cell models and curved quadratic cells mapped from triangular cells. The results are organized by the number of edges in model (the number of unknowns) and by the observer angle θ, for an observer position with $\phi = 0$.

as a sphere). The 108-edge model only employs 34 unknowns/λ^2, but because of the curved cells is able to produce similar accuracy for this problem with only 36% of the unknowns. Results at other angles in ϕ exhibit a comparable accuracy.

REFERENCES

[1] M. G. Duffy, "Quadrature over a pyramid or cube of integrands with a singularity at a vertex," *SIAM J. Num. Anal.*, vol. 19, pp. 1260–1262, 1982. doi:10.1137/0719090

[2] S. Järvenpää, M. Taskinen, and P. Ylä-Oijala, "Singularity extraction technique for integral equation methods with higher order basis functions on plane triangles and tetrahedra," *Int. J. Num. Meth. Eng.*, vol. 58, pp. 1149–1165, 2003. doi:10.1002/nme.810

[3] M. A. Khayat and D. R. Wilton, "Numerical evaluation of singular and near-singular potential integrals," *IEEE Trans. Antenn. Propagat.*, vol. 53, pp. 3180–3190. October 2005. doi:10.1109/TAP.2005.856342

[4] T. N. L. Patterson, "Generation of interpolatory quadrature rules of the highest degree of precision with preassigned nodes for general weight functions," *ACM Trans. Math Software*, vol. 15, pp. 137–143, June 1989. doi:10.1145/63522.69649

[5] R. Cools, "An encyclopaedia of cubature formulas," http://www.cs.kuleuven. ac.be/~nines/ecf/.

[6] A. F. Peterson, S. L. Ray, and R. Mittra, *Computational Methods for Electromagnetics*. New York: IEEE Press, 1998.

CHAPTER 7

CHAPTER 7

Use of Curl-conforming Bases with the Magnetic Field Integral Equation

As an illustration of the use of curl-conforming bases, the vector mapping procedures developed in Chapter 5 are applied to the magnetic field integral equation for scattering from closed perfectly conducting bodies with curved surfaces. The MFIE was derived in Section 1.1. The following sections discuss the process of discretizing the equation using the method of moments (MoM) procedure. Results are presented for scattering from conducting spheres, where they are compared to the exact solution and to results of the EFIE approach of Chapter 6.

7.1 TESTED FORM OF THE MFIE

As part of the MoM discretization process, the MFIE from Eq. (1.9) can be enforced by multiplication (by scalar product) with a vector testing function. This yields the equation

$$
\iint_{\text{surface}} \bar{T} \bullet \{\hat{n} \times \bar{H}^{\text{inc}}\} ds \, dt =
$$

$$
\iint_{\text{surface}} \bar{T} \bullet \bar{J} \, ds \, dt - \iint_{\text{surface}} \bar{T} \bullet \{\hat{n} \times \bar{H}^{s}\} ds \, dt
\tag{7.1}
$$

where the integration is performed over the surface of the conductor, using a "test-ing" function $\bar{T}(s,\ t)$ that is tangential to that surface. The vector $\hat{n}$ is the outward normal unit vector. The excitation function $\bar{H}^{\text{inc}}$ represents the magnetic field in-cident on the structure (in the absence of the structure) and is assumed known. The function $\bar{H}^s$ is the "scattered" magnetic field produced by the surface current $\bar{J}$, also in the structure's absence, which may be obtained from the expression

$$\bar{H}^s = \nabla \times \bar{A} \tag{7.2}$$

where

$$\bar{A}(s,\ t) = \iint_{\text{surface}} \bar{J}(s',\ t')G\,(R)ds'\,dt' \tag{7.3}$$

In (7.3), G is the free space Green's function

$$G(R) = \frac{e^{-jkR}}{4\pi\,R} \tag{7.4}$$

and

$$R = \sqrt{[x(s,\ t) - x(s',\ t')]^2 + [y(s,\ t) - y(s',\ t')]^2 + [z(s,\ t) - z(s',\ t')]^2} \tag{7.5}$$

When the source and observation locations coincide on the surface, R vanishes and the expressions in (7.1) and (7.3) are interpreted as the limiting cases when the observer approaches the surface from the exterior.

To provide motivation for using curl-conforming basis functions with (7.1), observe that the scattered magnetic field in (7.2) can be obtained from the equivalent expression

$$\bar{H}^s = \iiint \{\nabla' \times \bar{J}\}G\,ds'\,dt'\,dn' \tag{7.6}$$

In (7.6), despite $\bar{J}$ being confined to the surface, $\nabla \times \bar{J}$ must be interpreted as a three-dimensional generalized function (by including Dirac delta functions and their derivatives that arise when the three-dimensional curl operator is ap-plied to $\bar{J}$). While it is seldom convenient to use (7.6) to calculate the tangential field components on the surface, this expression *is* convenient for use if it is necessary to calculate the normal component of the magnetic field in the source region.

If curl-conforming basis functions are used for $\bar{J}$, (7.6) suggests that all components of the magnetic field will be well behaved (bounded, continuous) at cell boundaries on the surface. On the other hand, basis functions that do not impose cell-to-cell tangential-vector continuity result in an infinite normal component of the magnetic field at cell edges. We note that most numerical approaches for solving the MFIE have *not* used curl-conforming bases in the past, and it may not be essential to do so. However, curl-conforming basis functions appear to be appropriate for the MFIE operator.

7.2   ENTRIES OF THE MoM MATRIX

Suppose that the surface is subdivided into a mesh of curved cells based on a mapping of square or triangular reference cells. For instance, a mapping in terms of the quadratic Lagrangian interpolation polynomials of Sections 2.3 and 2.5 is convenient for illustration. The cell shapes are provided by explicit functions x, y, and z of local parameters u and v, as described in Chapter 2.

The MoM process requires the expansion of the surface current density in terms of basis functions. For illustration, curl-conforming vector bases $\{\bar{B}_n\}$ are distributed over the surface in accordance with the curved-cell model. The current is represented by the summation

$$\bar{J}(s,\ t) \cong \sum_{n=1}^{N} I_n \bar{B}_n(s,\ t) \tag{7.7}$$

where $\{I_n\}$ denote N complex-valued coefficients that henceforth are the unknowns to be determined. The curl-conforming functions are also employed as testing functions

$$\bar{T}_m(s,\ t) = \bar{B}_m(s,\ t) \tag{7.8}$$

to provide a means for obtaining N linearly independent equations from (7.1). The equations can be organized into matrix form to yield

$$\mathbf{H} = \mathbf{YI} \tag{7.9}$$

where $\mathbf{H}$ and $\mathbf{I}$ are $N \times 1$ column vectors and Y is an $N \times N$ matrix. The entries of $\mathbf{I}$ are the coefficients $\{I_n\}$, while those of $\mathbf{H}$ are given by

$$H_m = \iint\limits_{\text{surface}} \bar{T}_m \bullet \{\hat{n} \times \bar{H}^{\text{inc}}\} ds\, dt \tag{7.10}$$

The entries of $\mathbf{Y}$ are

$$Y_{mn} = \iint\limits_{\text{surface}} \bar{T}_m \bullet \bar{B}_n\, ds\, dt - \iint\limits_{\text{surface}} \bar{T}_m \bullet \{\hat{n} \times \bar{H}_n^s\} ds\, dt \tag{7.11}$$

where $\bar{H}_n^s$ denotes the magnetic field produced by the nth basis function. If the source region does not coincide with the observer location, the magnetic field can be easily obtained from the expression

$$\bar{H}_n^s = -\iint\limits_{\text{surface}} \bar{B}_n(s',\ t') \times \nabla G\, ds'\, dt' \tag{7.12}$$

where

$$\nabla G = -\frac{e^{-jkR}}{4\pi R^3}(1 + jkR)\{[x(s,\ t) - x(s',\ t')]\hat{x} \\ + [y(s,\ t) - y(s',\ t')]\hat{y} + [z(s,\ t) - z(s',\ t')]\hat{z}\} \tag{7.13}$$

When it is necessary to calculate $\bar{H}_n^s$ at a location within the source region, (7.12) must be evaluated in the limiting case from the exterior of the surface.

7.3   MAPPED MoM MATRIX ENTRIES

Since the surface is divided into cells, and each cell is defined by an independent mapping, the preceding integral expressions must be evaluated separately for each cell. Some or all of the basis and testing functions can be viewed as straddling cell pairs, so generally each matrix entry may involve integrals over more than one cell. Below, we consider the part of the matrix entry arising from a single observer cell (where the testing function is located) and a single source cell (where the basis function is located). The index (m or n) is used to denote specific testing and basis functions; the pointer arrays associated with the surface model can be used to identify the appropriate cells where these functions reside.

All the integrals required in Section 7.2 can be transformed into the local coordinates of each reference cell, as explained in Chapter 5. Here, explicit expressions are provided.

The curl-conforming basis functions on the curvilinear surface are defined according to the mapping from (5.68):

$$
\begin{bmatrix} B_x^{\text{curl}} \\ B_y^{\text{curl}} \\ B_z^{\text{curl}} \end{bmatrix} = \begin{bmatrix} \dfrac{\partial u}{\partial x} & \dfrac{\partial v}{\partial x} \\ \dfrac{\partial u}{\partial y} & \dfrac{\partial v}{\partial y} \\ \dfrac{\partial u}{\partial z} & \dfrac{\partial v}{\partial z} \end{bmatrix} \begin{bmatrix} R_u^{\text{curl}} \\ R_v^{\text{curl}} \end{bmatrix} = \mathbf{J}^{-1} \begin{bmatrix} R_u^{\text{curl}} \\ R_v^{\text{curl}} \end{bmatrix} \tag{7.14}
$$

In (7.14), $\bar{R}^{\text{curl}}$ denotes the basis function in the reference coordinates. (These functions may be normalized so that desired components have unity value at certain locations; for the moment we omit the normalization constant.) The testing functions are also to be curl-conforming functions, defined by a mapping of the same form as (7.14). As explained in Section 4.5, these curl-conforming functions are related to analogous divergence conforming functions by

$$
\hat{n} \times \bar{B}^{\text{div}} = \bar{B}^{\text{curl}} \tag{7.15}
$$

$$
\hat{n} \times \bar{B}^{\text{curl}} = -\bar{B}^{\text{div}} \tag{7.16}
$$

Because of the form of the MFIE in (7.1), it is convenient to exploit (7.15) and (7.16) as follows: Using the vector identity

$$
\bar{A} \bullet \bar{B} \times \bar{C} = \bar{A} \times \bar{B} \bullet \bar{C} \tag{7.17}
$$

Equation (7.10) can be rewritten in the form

$$
\begin{aligned}
H_m &= \iint_{\text{surface}} \bar{T}_m^{\text{curl}} \bullet \{\hat{n} \times \bar{H}^{\text{inc}}\} \, ds \, dt \\
&= \iint_{\text{surface}} \{\bar{T}_m^{\text{curl}} \times \hat{n}\} \bullet \bar{H}^{\text{inc}} \, ds \, dt \\
&= \iint_{\text{surface}} \bar{T}_m^{\text{div}} \bullet \bar{H}^{\text{inc}} \, ds \, dt
\end{aligned}
\tag{7.18}
$$

In (7.18), $\bar{T}_m^{\text{div}}$ denotes a divergence-conforming test function related to the specific curl-conforming test function under consideration by

$$
\bar{T}_m^{\text{div}} = -\hat{n} \times \bar{T}_m^{\text{curl}}
\tag{7.19}
$$

In a similar manner, the second integral in (7.11) may be recast as

$$
\iint_{\text{surface}} \bar{T}_m^{\text{curl}} \bullet \{\hat{n} \times \bar{H}_n^s\} \, ds \, dt = \iint_{\text{surface}} \bar{T}_m^{\text{div}} \bullet \bar{H}_n^s \, ds \, dt
\tag{7.20}
$$

The divergence-conforming functions are defined using the mapping of (5.77):

$$
\begin{bmatrix} T_x^{\text{div}} \\ T_y^{\text{div}} \\ T_z^{\text{div}} \end{bmatrix} = \frac{1}{D} \begin{bmatrix} \dfrac{\partial x}{\partial u} & \dfrac{\partial x}{\partial v} \\ \dfrac{\partial y}{\partial u} & \dfrac{\partial y}{\partial v} \\ \dfrac{\partial z}{\partial u} & \dfrac{\partial z}{\partial v} \end{bmatrix} \begin{bmatrix} R_u^{\text{div}} \\ R_v^{\text{div}} \end{bmatrix} = \frac{1}{D} \mathbf{J}^T \begin{bmatrix} R_u^{\text{div}} \\ R_v^{\text{div}} \end{bmatrix}
\tag{7.21}
$$

where $\bar{R}^{\text{div}}$ denotes the testing function in the reference coordinates. (The normalization of the testing functions, as described in Section 5.6, is optional.) In order to construct the scalar vector products arising within (7.11), (7.18), and (7.20) by means of matrix manipulations, we observe that

$$
\begin{bmatrix} T_x^{\text{div}} \\ T_y^{\text{div}} \\ T_z^{\text{div}} \end{bmatrix}^{\text{T}} = \frac{1}{D} \begin{bmatrix} R_u^{\text{div}} & R_v^{\text{div}} \end{bmatrix} \mathbf{J}
\tag{7.22}
$$

and

$$\begin{bmatrix} T_x^{\text{curl}} \\ T_y^{\text{curl}} \\ T_z^{\text{curl}} \end{bmatrix}^{\text{T}} = \begin{bmatrix} R_u^{\text{curl}} & R_v^{\text{curl}} \end{bmatrix} \mathbf{J}^{-\text{T}} \tag{7.23}$$

Using (7.22), the integral in (7.18) may be expressed as

$$H_m = \iint_{\text{cell on surface}} \bar{T}_m^{\text{div}} \bullet \bar{H}^{\text{inc}} \, ds \, dt = \iint_{\text{reference cell}} \begin{bmatrix} R_u^{\text{div}} & R_v^{\text{div}} \end{bmatrix}_m \mathbf{J} \begin{bmatrix} H_x^{\text{inc}} \\ H_y^{\text{inc}} \\ H_z^{\text{inc}} \end{bmatrix} du \, dv \tag{7.24}$$

where the matrix $\mathbf{J}$ for the cell is defined in (5.53), and where we also used the differential surface area

$$ds \, dt = D \, du \, dv \tag{7.25}$$

The scale factors $1/D$ from (7.22) and D from (7.25) cancel in the integrand of (7.24). This integral is to be evaluated by numerical quadrature carried out in the reference coordinates u and v, with the incident field sampled at the desired quadrature points with the assistance of the mapping functions $x(u, v)$, $y(u, v)$, and $z(u, v)$.

The first integral in (7.11) can be expressed over a single cell, using (7.23) and (7.25), as

$$\iint_{\text{cell on surface}} \bar{T}_m^{\text{curl}} \bullet \bar{B}_n^{\text{curl}} \, ds \, dt = \iint_{\text{reference cell}} \begin{bmatrix} R_u^{\text{curl}} & R_v^{\text{curl}} \end{bmatrix}_m \mathbf{J}^{-\text{T}} \mathbf{J}^{-1} \begin{bmatrix} R_u^{\text{curl}} \\ R_v^{\text{curl}} \end{bmatrix}_n D \, du \, dv \tag{7.26}$$

The entries of the 3×2 matrix $\mathbf{J}^{-1}$, defined in (5.54), are not explicitly provided by the mapping functions $x(u, v)$, etc. These entries are obtained in practice by inverting the numerical 3×3 Jacobian matrix in Eq. (5.64) at each of the necessary quadrature points.

Using (7.20), the second integral in (7.11) can be written as

$$\iint_{\text{cell on surface}} \bar{T}_m^{\text{curl}} \bullet \{\hat{n} \times \bar{H}_n^s\} ds \, dt = \iint_{\text{cell on surface}} \bar{T}_m^{\text{div}} \bullet \bar{H}_n^s \, ds \, dt$$

$$= \iint_{\text{reference cell (observer)}} \left[R_u^{\text{div}} \; R_v^{\text{div}} \right]_m \mathbf{J}|_{\text{observer}} \begin{bmatrix} H_x^s \\ H_y^s \\ H_z^s \end{bmatrix} du \, dv \tag{7.27}$$

where, using (7.12),

$$\begin{bmatrix} H_x^s \\ H_y^s \\ H_z^s \end{bmatrix} = - \iint_{\text{reference cell (source)}} \begin{bmatrix} 0 & \frac{\partial G}{\partial z} & -\frac{\partial G}{\partial y} \\ -\frac{\partial G}{\partial z} & 0 & \frac{\partial G}{\partial x} \\ \frac{\partial G}{\partial y} & -\frac{\partial G}{\partial x} & 0 \end{bmatrix} \mathbf{J}^{-1}|_{\text{source}} \begin{bmatrix} R_u^{\text{curl}} \\ R_v^{\text{curl}} \end{bmatrix}_n D \, du \, dv$$

$$\tag{7.28}$$

The integral in (7.27) is carried out over the observer cell, which is transformed to a reference cell using the Jacobian matrix $\mathbf{J}|_{\text{observer}}$ associated with that specific mapping. (As in (7.24), the scale factors $1/D$ and D cancel.) Equation (7.28) is embedded within (7.27), and involves an integral over the source cell, which is usually different from the observer cell. Since the cells are different, the mapping is different, and the transformation in (7.28) uses a matrix $\mathbf{J}^{-1}|_{\text{source}}$ that is not related to $\mathbf{J}|_{\text{observer}}$. These integrals must also be evaluated in the reference coordinates by numerical quadrature, with the derivatives of G in (7.28) determined from (7.13), and the entries of $\mathbf{J}^{-1}|_{\text{source}}$ obtained by the explicit numerical inversion of (5.64) at the quadrature points. The mapping functions $\{x(u, v), y(u, v), z(u, v)\}$ for both the source cell and the observer cell are required to compute the arguments of the derivatives of G. When the source and observer cells coincide, the quadrature procedure must be modified to handle the singularity in G as explained in Section 7.5.

7.4   NORMALIZATION OF CURL-CONFORMING BASIS FUNCTIONS

If interpolatory basis functions are used, and it is desired to interpret the coefficients $\{I_n\}$ as values of the surface current density, the curl-conforming basis functions used for $\bar{J}$ must be normalized so that the tangential component of each $\bar{B}_n$ has unity value at an appropriate location within each cell (in the curvilinear x–y–z space). This is accomplished by scaling each basis function by a constant equal to the magnitude of the appropriate base vector at that location, as explained in Section 5.4. For functions that straddle two cells (any of the "edge-based" functions of Chapter 4), the normalization constant may also contain a sign that ensures that the basis function points in a consistent tangential-vector direction from a global perspective. The normalization is incorporated into the matrix entries of the preceding section by multiplying the various components of the basis functions in the reference cell by the appropriate constants. These constants are cell-specific, and can be determined at the start of the analysis and stored in an array for convenient reference as required during the matrix construction phase of the MoM procedure, and any post-processing that involves the basis functions.

7.5   TREATMENT OF THE SINGULARITY OF THE GREEN'S FUNCTION

When the source and observer cells in (7.27) and (7.28) coincide, there will generally be points where the function R within G vanishes. In fact, the integral involves a singularity that must be treated with care.

For an observation point (s, t) located on a smooth portion of the source cell, the singularity can be treated in an analytical manner by replacing Eq. (7.12) with

$$\bar{H}_n^s(s,\ t) = \frac{\bar{B}_n(s,\ t)}{2} - \iint\limits_{surface-\varepsilon} \bar{B}_n(s',\ t') \times \nabla G \, ds' \, dt' \tag{7.29}$$

where the remaining integral is to exclude an infinitesimal region around the point (s, t) from the domain of integration [1]. The primary impact of this operation is the addition of the leading term on the right-hand side of (7.29), which effectively changes the first integral in (7.11) to one having only half its value:

$$Y_{mn} = \frac{1}{2} \iint\limits_{surface} \bar{T}_m \bullet \bar{B}_n \, ds \, dt - \iint\limits_{surface} \bar{T}_m \bullet \{\hat{n} \times \bar{H}_n^s\} ds \, dt \tag{7.30}$$

The calculation of $\bar{H}_n^s$ in the second term in (7.30) is modified in principle by the exclusion of an ε-neighborhood of the observation point from the domain.

However, in practice when evaluating this expression with numerical quadrature, as long as a node of the quadrature rule does not coincide with (s, t) the effect is the same whether or not the neighborhood is excluded.

Even after the exclusion of the ε-neighborhood of (s, t), the integrand in (7.29) still behaves as $1/R$ and will be difficult to evaluate accurately with standard quadrature rules. The $1/R$ behavior can be dealt with using the Duffy transformation approach described in Chapter 6. The reference cell can be divided into three or four triangular cells, each with a corner at the singularity, and the integrals over each cell can be evaluated after a change of variable that cancels the singularity as explained in Section 6.5. The result is that three or four integrals must be evaluated over square domains, and their results combined, to produce the result of the integral in (7.29) for each location of (s, t). Alternatives to the Duffy transformation are cited in Chapter 6.

7.6   RESULTS

A computer program was developed that implements the preceding expressions for the MoM matrix entries associated with $p = 0$ basis functions from Chapter 4 and quadratic cells mapped from triangular reference cells. For illustration, consider a perfectly conducting sphere of radius 0.5λ, where λ denotes the wavelength. Figure 7.1 depicts the current magnitude $|J_\theta|$ produced by a uniform plane wave incident on a sphere, obtained from the MFIE procedure with a model containing 300 cell edges (equivalently, 300 unknowns). The representation for J_θ along this cut is piecewise constant, as explicitly depicted in the figure. The sphere model is described in Section 6.7. The exact solution is shown for comparison.

Table 7.1 shows the percentage error in the current density, for several models of the sphere of radius 0.5λ. The MFIE performance is compared with that of the EFIE approach from Chapter 6. The error was determined by averaging over the difference between the current density coefficients and the exact solution for the appropriate current component (normal to an edge for the EFIE, tangential to an

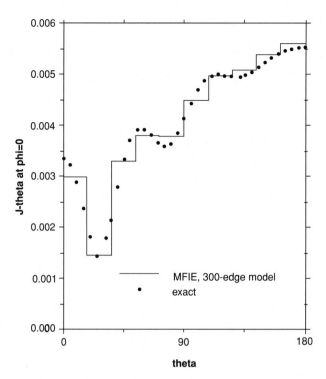

FIGURE 7.1: Magnitude of J_θ along the $\phi = 0$ cut, produced by a plane wave incident in the $\theta = 0$ direction with an x-component of electric field. The numerical solution obtained from the MFIE with curl-conforming basis functions is compared with the exact solution.

TABLE 7.1: Average Percent Error in the Current Density

NO. OF EDGES IN MODEL	LARGEST EDGE LENGTH IN MODEL (λ)	EFIE (DIV-CONF.), %	MFIE (CURL-CONF.), %
108	0.36	6.2	5.8
192	0.29	3.5	3.8
300	0.24	2.3	2.8
432	0.20	1.7	2.1

Note. Averaged over all edges at the edge center, for a sphere of radius 0.5λ modeled with curved quadratic cells mapped from triangles.

edge for the MFIE) at the edge centers. The error levels for the EFIE and MFIE decrease as the models are refined, and are similar for the two approaches.

REFERENCES

[1]   A. J. Poggio and E. K. Miller, "Integral equation solutions of three-dimensional scattering problems," in *Computer Techniques for Electromagnetics*, R. Mittra, Ed. New York: Hemisphere, 1987.

Biography

Andrew F. Peterson received the B.S., M.S., and Ph.D. degrees in electrical engineering from the University of Illinois, Urbana-Champaign in 1982, 1983, and 1986, respectively. Since 1989, he has been a member of the faculty of the School of Electrical and Computer Engineering at the Georgia Institute of Technology, where he is now Professor and Associate Chair for Faculty Development. He teaches electromagnetic field theory and computational electromagnetics, and conducts research in the development of computational techniques for electromagnetic scattering, microwave devices, and electronic packaging applications. He is the principal author of the text *Computational Methods for Electromagnetics*, IEEE Press, 1998.

Dr. Peterson is a past recipient of the ONR Graduate Fellowship and the NSF Young Investigator Award. He has served as an Associate Editor of the *IEEE Transactions on Antennas and Propagation*, as an Associate Editor of the *IEEE Antennas and Wireless Propagation Letters*, as the General Chair of the 1998 IEEE AP-S International Symposium and URSI/USNC Radio Science Meeting, and as a member of IEEE AP-S AdCom. He also served for 6 years as a Director of ACES, and 2 years as Chair of the IEEE Atlanta Section. He will serve as the President of the IEEE AP-S during 2006. He is an IEEE Fellow and a recipient of the IEEE Third Millennium Medal. He is also a member of the Applied Computational Electromagnetics Society (ACES), the International Union of Radio Scientists (URSI) Commission B, the American Society for Engineering Education, and the American Association of University Professors.